P9-ARP-213

Lecture Notes in Mathematics

For information about Vols. 1–1201 please contact your bookseller or Springer-Verlag

Vol. 1202: A. Dür, Möbius Functions, Incidence Algebras and Power Series Representations. XI, 134 pages. 1986.

Vol. 1203: Stochastic Processes and Their Applications. Proceedings, 1985. Edited by K. Itô and T. Hida. VI, 222 pages. 1986.

Vol. 1204: Séminaire de Probabilités XX, 1984/85. Proceedings. Edité par J. Azéma et M. Yor. V, 639 pages. 1986.

Vol. 1205: B.Z. Moroz, Analytic Arithmetic in Algebraic Number Fields. VII, 177 pages. 1986.

Vol. 1206: Probability and Analysis, Varenna (Como) 1985. Seminar. Edited by G. Letta and M. Pratelli. VIII, 280 pages. 1986.

Vol. 1207: P.H. Bérard, Spectral Geometry: Direct and Inverse Problems. With an Appendix by G. Besson. XIII, 272 pages. 1986.

Vol. 1208: S. Kaijser, J.W. Pelletier, Interpolation Functors and Duality. IV, 167 pages. 1986.

Vol. 1209: Differential Geometry, Peñíscola 1985. Proceedings. Edited by A.M. Naveira, A. Ferrández and F. Mascaró. VIII, 306 pages. 1986.

Vol. 1210: Probability Measures on Groups VIII. Proceedings, 1985. Edited by H. Heyer. X, 386 pages. 1986.

Vol. 1211: M.B. Sevryuk, Reversible Systems. V, 319 pages. 1986.

Vol. 1212: Stochastic Spatial Processes. Proceedings, 1984. Edited by P. Tautu. VIII, 311 pages. 1986.

Vol. 1213: L.G. Lewis, Jr., J.P. May, M. Steinberger, Equivariant Stable Homotopy Theory. IX, 538 pages. 1986.

Vol. 1214: Global Analysis – Studies and Applications II. Edited by Yu.G. Borisovich and Yu.E. Gliklikh. V, 275 pages. 1986.

Vol. 1215: Lectures in Probability and Statistics. Edited by G. del Pino and R. Rebolledo. V, 491 pages. 1986.

Vol. 1216: J. Kogan, Bifurcation of Extremals in Optimal Control. VIII, 106 pages. 1986.

Vol. 1217: Transformation Groups. Proceedings, 1985. Edited by S. Jackowski and K. Pawalowski. X, 396 pages. 1986.

Vol. 1218: Schrödinger Operators, Aarhus 1985. Seminar. Edited by E. Balslev. V, 222 pages. 1986.

Vol. 1219: R. Weissauer, Stabile Modulformen und Eisensteinreihen. III, 147 Seiten. 1986.

Vol. 1220: Séminaire d'Algèbre Paul Dubreil et Marie-Paule Malliavin. Proceedings, 1985. Edité par M.-P. Malliavin. IV, 200 pages. 1986.

Vol. 1221: Probability and Banach Spaces. Proceedings, 1985. Edited by J. Bastero and M. San Miguel. XI, 222 pages. 1986.

Vol. 1222: A. Katok, J.-M. Strelcyn, with the collaboration of F. Ledrappier and F. Przytycki, Invariant Manifolds, Entropy and Billiards; Smooth Maps with Singularities. VIII, 283 pages. 1986.

Vol. 1223: Differential Equations in Banach Spaces. Proceedings, 1985. Edited by A. Favini and E. Obrecht. VIII, 299 pages. 1986.

Vol. 1224: Nonlinear Diffusion Problems, Montecatini Terme 1985. Seminar. Edited by A. Fasano and M. Primicerio. VIII, 188 pages. 1986.

Vol. 1225: Inverse Problems, Montecatini Terme 1986. Seminar. Edited by G. Talenti. VIII, 204 pages. 1986.

Vol. 1226: A. Buium, Differential Function Fields and Moduli of Algebraic Varieties. IX, 146 pages. 1986.

Vol. 1227: H. Helson, The Spectral Theorem. VI, 104 pages. 1986.

Vol. 1228: Multigrid Methods II. Proceedings, 1985. Edited by W. Hackbusch and U. Trottenberg. VI, 336 pages. 1986.

Vol. 1229: O. Bratteli, Derivations, Dissipations and Group Actions on C*-algebras. IV, 277 pages. 1986.

Vol. 1230: Numerical Analysis. Proceedings, 1984. Edited by J.-P. Hennart. X, 234 pages. 1986.

Vol. 1231: E.-U. Gekeler, Drinfeld Modular Curves. XIV, 107 pages. 1986.

Vol. 1232: P.C. Schuur, Asymptotic Analysis of Soliton Problems. VIII, 180 pages. 1986.

Vol. 1233: Stability Problems for Stochastic Models. Proceedings, 1985. Edited by V.V. Kalashnikov, B. Penkov and V.M. Zolotarev. VI, 223 pages. 1986.

Vol. 1234: Combinatoire énumérative. Proceedings, 1985. Edité par G. Labelle et P. Leroux. XIV, 387 pages. 1986.

Vol. 1235: Séminaire de Théorie du Potentiel, Paris, No. 8. Directeurs: M. Brelot, G. Choquet et J. Deny. Rédacteurs: F. Hirsch et G. Mokobodzki. III, 209 pages. 1987.

Vol. 1236: Stochastic Partial Differential Equations and Applications. Proceedings, 1985. Edited by G. Da Prato and L. Tubaro. V, 257 pages. 1987.

Vol. 1237: Rational Approximation and its Applications in Mathematics and Physics. Proceedings, 1985. Edited by J. Gilewicz, M. Pindor and W. Siemaszko. XII, 350 pages. 1987.

Vol. 1238: M. Holz, K.-P. Podewski and K. Steffens, Injective Choice Functions. VI, 183 pages. 1987.

Vol. 1239: P. Vojta, Diophantine Approximations and Value Distribution Theory. X, 132 pages. 1987.

Vol. 1240: Number Theory, New York 1984–85. Seminar. Edited by D.V. Chudnovsky, G.V. Chudnovsky, H. Cohn and M.B. Nathanson. V, 324 pages. 1987.

Vol. 1241: L. Gårding, Singularities in Linear Wave Propagation. III, 125 pages. 1987.

Vol. 1242: Functional Analysis II, with Contributions by J. Hoffmann-Jørgensen et al. Edited by S. Kurepa, H. Kraljević and D. Butković. VII, 432 pages. 1987.

Vol. 1243: Non Commutative Harmonic Analysis and Lie Groups. Proceedings, 1985. Edited by J. Carmona, P. Delorme and M. Vergne. V, 309 pages. 1987.

Vol. 1244: W. Müller, Manifolds with Cusps of Rank One. XI, 158 pages. 1987.

Vol. 1245: S. Rallis, L-Functions and the Oscillator Representation. XVI, 239 pages. 1987.

Vol. 1246: Hodge Theory. Proceedings, 1985. Edited by E. Cattani, F. Guillén, A. Kaplan and F. Puerta. VII, 175 pages. 1987.

Vol. 1247: Séminaire de Probabilités XXI. Proceedings. Edité par J. Azéma, P.A. Meyer et M. Yor. IV, 579 pages. 1987.

Vol. 1248: Nonlinear Semigroups, Partial Differential Equations and Attractors. Proceedings, 1985. Edited by T.L. Gill and W.W. Zachary. IX, 185 pages. 1987.

Vol. 1249: I. van den Berg, Nonstandard Asymptotic Analysis. IX, 187 pages. 1987.

Vol. 1250: Stochastic Processes – Mathematics and Physics II. Proceedings 1985. Edited by S. Albeverio, Ph. Blanchard and L. Streit. VI, 359 pages. 1987.

Vol. 1251: Differential Geometric Methods in Mathematical Physics. Proceedings, 1985. Edited by P.L. García and A. Pérez-Rendón. VII, 300 pages. 1987.

Vol. 1252: T. Kaise, Représentations de Weil et GL_2 Algèbres de division et GL_n. VII, 203 pages. 1987.

Vol. 1253: J. Fischer, An Approach to the Selberg Trace Formula via the Selberg Zeta-Function. III, 184 pages. 1987.

Vol. 1254: S. Gelbart, I. Piatetski-Shapiro, S. Rallis. Explicit Constructions of Automorphic L-Functions. VI, 152 pages. 1987.

Vol. 1255: Differential Geometry and Differential Equations. Proceedings, 1985. Edited by C. Gu, M. Berger and R.L. Bryant. XII, 243 pages. 1987.

Vol. 1256: Pseudo-Differential Operators. Proceedings, 1986. Edited by H.O. Cordes, B. Gramsch and H. Widom. X, 479 pages. 1987.

Vol. 1257: X. Wang, On the C*-Algebras of Foliations in the Plane. V, 165 pages. 1987.

Vol. 1258: J. Weidmann, Spectral Theory of Ordinary Differential Operators. VI, 303 pages. 1987.

continued on page 113

Lecture Notes in Mathematics

Edited by A. Dold, B. Eckmann and F. Takens

1422

V. Villani (Ed.)

Complex Geometry and Analysis

Proceedings of the International Symposium
in honour of Edoardo Vesentini
held in Pisa (Italy), May 23–27, 1988

Springer-Verlag
Berlin Heidelberg New York London Paris Tokyo Hong Kong

Editor

Vinicio Villani
Dipartimento di Matematica, Università di Pisa
Via Buonarroti, 2, 56100 Pisa, Italy

Mathematics Subject Classification (1980): 32-xx, 53-xx, 47-xx

ISBN 3-540-52434-7 Springer-Verlag Berlin Heidelberg New York
ISBN 0-387-52434-7 Springer-Verlag New York Berlin Heidelberg

Printed in Germany

Printing and binding: Druckhaus Beltz, Hemsbach/Bergstr.
2146/3140-543210 – Printed on acid-free paper

Foreword

This volume contains the texts of the main talks delivered at the International Symposium on Complex Geometry and Analysis held in Pisa, May 23–27, 1988. The Symposium was organized on the occasion of the sixtieth birthday of Edoardo Vesentini, by some of his former students, in appreciation of his many contributions to mathematics, of his teaching and advice.

The aim of the lectures was to describe the present situation, the recent developments and research trends in several relevant topics in Complex Geometry and Analysis, that is in those fields in which the mathematical activity of E. Vesentini is most fruitful and inspiring. The contributors are distinguished mathematicians who have actively collaborated with the mathematical school in Pisa over the past thirty years.

The organizers would like to thank all the supporting institutions, and, in particular, the Comitato per la Matematica (CNR) and the Gruppo Nazionale di Geometria Analitica ed Analisi Complessa (MPI).

The Organizing Committee
V. Villani (chairman)
T. Franzoni
G. Gentili
G. Gigante
S. Levi
F. Ricci
G. Tomassini

Contents

HYPERKÄHLER MANIFOLDS

Michael Atiyah
Mathematical Institute,
24-29 St. Giles,
Oxford OX1 3LB,
England, U.K.

§1. Introduction and Definitions

In recent years hyperkähler manifolds have turned up in a wide variety of contexts, and it is now becoming clear that they form a very interesting class of manifolds with a rich theory. The purpose of this lecture is to justify these claims by giving an overall survey of the field.

I shall begin by reviewing the basic definitions and elementary properties. Then in §2 I will describe the hyperkähler quotient construction of [7] which enables us to construct many examples quite painlessly. This shows that the theory has a definitely non-trivial content. In §3 I will concentrate on 4-dimensional manifolds which are of special interest for various reasons including the classical relation to physics via Einstein's equations. In particular, I will describe the beautiful family of examples due to P.B. Kronheimer [9]. In §4 I explain how Yang-Mills moduli spaces give yet more examples of hyperkähler manifolds. In particular the moduli spaces of magnetic monopoles studied in §2 are of special interest. Finally in §5 I describe the twistor theory of R. Penrose as it applies to hyperkähler manifolds and illustrate it for the case of monopole spaces.

As the name rather obviously suggests hyperkähler manifolds are a generalization of Kähler manifolds, so it is best to start by briefly recalling that a Kähler manifold may be defined as a <u>Riemannian</u> manifold X with an <u>almost complex structure</u> I (orthogonal transformation of the tangent bundle with $I^2 = -1$) which is <u>covariant constant</u>. This condition implies the usual integrability condition for I so that X is actually a <u>complex manifold</u>. Equivalently a Kähler manifold is a Riemannian manifold with holonomy group contained in

$$U(n) \subset SO(2n) .$$

The importance of Kähler manifolds lies mainly in the fact that algebraic manifolds (affine or projective) always carry Kähler metrics. It is also significant that the 2-form ω_I associated to I and the metric is closed and non-degenerate, so that Kähler manifolds are in particular symplectic.

Next let us recall that the algebra H of quaternions is generated (over R) by the symbols i, j, k with the relations

$$i^2 = j^2 = k^2 = -1$$

$$ij = -ji = k \quad \text{etc.}$$

or more succinctly, for $x,y,z \in R$,

$$(xi + yj + zk)^2 = -(x^2 + y^2 + z^2) \tag{1.1}$$

A hyperkähler manifold is now defined as a Riemannian manifold X endowed with I, J, K (orthogonal transformations of the tangent bundle) satisfying the quaternion algebra identities and covariant constant. Briefly we may say that the tangent spaces to X have a covariant constant H-module structure. Equivalently the holonomy group of X lies in the symplectic group

$$Sp(k) \subset SO(4k) .$$

Clearly by choosing the structure I and ignoring J, K we see that X has in particular a complex Kähler structure. More generally the role of I can be replaced by

$$I_\lambda = xI + yJ + zK$$

where

$$\lambda = (x,y,z) \in R^3 \quad \text{with} \quad \lambda^2 = x^2 + y^2 + z^2 = 1 .$$

This shows that X has a whole family of complex structures, parametrized by points λ of the 2-sphere, and that the metric is

Kählerian for all these complex structures. This explains the terminology "hyperkähler".

Note. Although logical and descriptive the terminology is rather cumbersome and a beautiful class of manifolds deserves a better fate. Because they involve so many of Hamilton's main interests (quaternions, symplectic geometry, theoretical physics) I proposed at one stage that they should be christened "Hamiltonian manifolds". Unfortunately the hyperkähler usage was too widely used (expecially by physicists) to be eradicated. A pity!

The 3 operators I, J, K combined with the metric yield 3 covariant constant 2-forms ω_I, ω_J, ω_K giving 3 symplectic structures. If we fix on the complex structure defined by I then ω_I is the $(1,1)$ form associated to the Kähler metric, while $\omega_J + i\omega_K$ becomes a closed holomorphic 2-form defining a "holomorphic symplectic" structure.

Since Kähler manifolds play an important role in complex algebraic geometry one might speculate that hyperkähler manifolds should play a similarly important role in "quaternionic algebraic geometry". Unfortunately quaternionic algebraic geometry does not seem to exist. In fact, as we shall see, one can argue in reverse that hyperkähler geometry provides a substitute for the non-existent quaternionic algebraic geometry. The arguments for this view-point are strengthened by the following important fact: an irreducible hyperkähler metric is uniquely determined (up to a constant scale factor) by its family of complex structures. Note that the corresponding result for Kähler metrics is totally false: there are many Kähler metrics on a fixed complex manifold. In this sense hyperkähler geometry is more tightly related to complex analysis (and eventually to algebra) than Kähler geometry. This becomes clearer in the twistor picture which we shall explain in §5.

Clearly quaternionic space H^k with standard metric (where i, j, k act orthogonally) is a hyperkähler manifold. These linear or flat examples are not very interesting, but they provide the starting point for the construction of non-linear examples as we shall see in the next section.

§2. The quotient construction

It will be clear from the definitions in §1 that a hyperkähler structure is a very restricted one, and one might tend to dismiss the theory as having only a mild specialized interest. That was certainly my initial reaction, but my view was radically changed by the discovery in [7] of a very simple and beautiful "quotient construction" which generates vast numbers of hyperkähler manifolds in a natural way. Moreover this quotient construction is the quaternionic analogue of a Kähler quotient which is the geometric version of classical invariant theory. In this sense the hyperkähler quotient replaces the non-existent "quaternionic invariant theory".

Let me begin therefore by reviewing the quotient construction in Kähler geometry. The prototype is provided by considering the standard action (scalar multiplication) of the circle group S^1 on a complex vector space C^n. The standard way in algebraic geometry to form a quotient is to complexify S^1 to the complex multiplicative group C^*, then remove the origin from C^n (a "bad" point) and to form the projective space

$$P_{n-1} = (C^n - 0)/C^* . \tag{2.1}$$

An equivalent procedure using real differential geometry is to restrict the action of S^1 to the unit sphere S^{2n-1}, so that

$$P_{n-1} = S^{2n-1}/S^1 . \tag{2.2}$$

In this guise P_{n-1} inherits a natural metric, but the complex structure is not so transparant. The link between the complex and metric view-points lies in symplectic geometry. In fact the function $|z|^2$ on C^n viewed as a Hamiltonian, with respect to the symplectic structure of C^n given by its standard hermitian metric, generates the Hamiltonian flow of the S^1-action. The quotient (2.2) inherits a natural symplectic structure, a procedure well known in classicial mechanics.

This simple example generalizes to the action of any compact (connected) Lie group G on a Kähler manifold X. We assume that G preserves both the metric and the complex structure, hence also the symplectic structure. Under mild conditions there is then a <u>moment map</u>

$$\mu : X \to g^* \tag{2.3}$$

where g^* is the dual of the Lie algebra of G. The components of μ are Hamiltonian functions corresponding to the flows defined by one-parameter subgroups of G. Also μ is assumed to be G-equivariant. Now let $\alpha \in g^*$ be fixed by G (frequently we take $\alpha = 0$) and assume it is a regular value for μ. Then the manifold

$$X_\alpha = \mu^{-1}(\alpha)/G$$

inherits a natural symplectic structure. Clearly X_α also inherits a Riemannian metric. Together with the symplectic form ω this then defines an almost complex structure I which makes X_α a <u>Kähler manifold</u>, the <u>Kähler quotient</u> [8].

If X is a projective algebraic variety (with Kähler class coming from a projective embedding) then X_α is the projective variety whose coordinate ring is essentially the G-invariant part of the coordinate ring of X. All this is part of "geometric invariant theory" as developed by Mumford.

We are now ready for the hyperkähler case, so let X be a hyperkähler manifold and let G be a compact Lie group of automorphisms of X. Using the 3 symplectic structures ω_I, ω_J, ω_K of X we get (under mild assumptions) 3 moment maps μ_I, μ_J, μ_K which we can combine into a single quaternionic moment map

$$\mu : X \to g^* \otimes R^3$$

which is G-equivariant. Let $\alpha \in g^* \otimes R^3$ be fixed by G and assume this is a regular value of μ. Then the manifold

$$X_\alpha = \mu^{-1}(\alpha)/G$$

has 3 induced symplectic structures which, together with the induced metric, define a hyperkähler structure. This is the <u>hyperkähler quotient</u> of [7].

The complex structure I of X_α can be seen from an alternative description. Recall that $\omega_J + i\omega_K$ defines a holomorphic symplectic structure on X. The holomorphic action of G^c preserves this and $\mu_J + i\mu_K$ gives a holomorphic moment map μ^c. Then $(\mu^c)^{-1}(\alpha_J + i\alpha_K)$ is a complex submanifold Y of X acted on by G and X_α is clearly the same as the Kähler quotient Y_{α_I}, where $\alpha_I, \alpha_J, \alpha_K \in g^*$ are the 3 components of $\alpha \in g^* \otimes R^3$.

Notes. 1) As the title of [7] indicates hyperkähler manifolds are of interest to physicists in relation to supersymmetric models.

2) If X is complete the hyperkähler quotient X_α is also complete. If α is not a regular value of μ then X_α will have singularities and removing these leads to an incomplete manifold.

As I pointed out in §1 the quaternionic spaces H^k are hyperkähler manifolds. Hence if $G \to Sp(k) = Aut(H^k)$ is any symplectic representation of G we are in the situation where we can try to construct quotient hyperkähler manifolds from the action of G on H^k. Since there are many choices of groups and representation we see that the quotient construction will lead to very many hyperkähler manifolds. Even when G is a circle or torus the construction yields interesting examples.

§3. 4-dimensional examples

Since a hyperkähler manifold has dimension $4k$ the lowest dimension is 4, i.e. quaternionic dimension 1. These are in a sense the quaternionic analogues of Riemann surfaces or algebraic curves and, as such, deserve special attention. They are also of special interest because 4 is the dimension of space-time and, since $Sp(1) = SU(2)$, a hyperkähler 4-manifold is the same as a Kähler-Einstein (or self-dual Einstein) manifold. As solutions of the (positive definite) Einstein equations such manifolds have been studied by pysicists in connection with the quantization of gravity. They are referred to as "gravitational instantons".

So far I have not discussed questions of compactness or completeness but these are obviously important aspects. Compact 4-dimensional examples are scarce and essentially consist of flat tori and the K3 surfaces where the existence of a Kähler-Einstein metric has been established by S.T. Yau with his proof of the Calabi conjecture.

If we consider non-compact manifolds the next simplest class would be complete manifolds which are asymptotically flat. In fact this can be interpreted in a number of slightly different ways. One class (referred to as ALE spaces: asymptotically locally Euclidean) requires the manifold to behave at ∞ like $(R^4-0)/\Gamma$ where $\Gamma \subset Sp(1)$ is a finite group.

Since $Sp(1) = SU(2)$ double covers $SO(3)$ the groups Γ which can occur are just the double covers of the symmetry groups of the Platonic regular solids in R^3, namely the cyclic, dihedral, tetrahedral, octahedral and icosahedral groups. These groups are well-known to be linked, in a subtle way, to the simply-laced Lie groups A_n, D_n, E_6, E_7, E_8.

The construction and classification of ALE spaces for all choices of Γ has been worked out by P.B. Kronheimer [9] in a very beautiful theory. He constructs his manifolds as hyperkähler quotients with a judicious choice of Lie group G and symplectic representation. These are determined uniformly, for all Γ, in terms of the representation theory of Γ, the key ingredients being the regular representation and the 2-dimensional representation $\Gamma \to Sp(1) = SU(2)$ from which Γ arose. Moreover, the hyperkähler metrics have moduli which arise from the choice of α for the value of the moment map. Kronheimer proves that the moduli space can be naturally identified with an open set of "regular" points in the quotient

$$(h \otimes R^3)/W$$

where h is the Cartan algebra of the corresponding Lie group and W is its Weyl group.

If we consider these 4-dimensional hyperkähler manifolds as "quaternionic algebraic curves" they are analogous in many respects to complex algebraic curves. H^1 of complex curves is replaced by H^2 of our "quaternionic curves" so that

$$\text{rank } H^2 \ (= \dim h)$$

is analogous to the genus. The most direct analogy would restrict us to the cyclic groups (type A_n) but the quaternionic case is richer since we have another infinite family (D_n) and the 3 exceptional cases. Also the moduli are determined by period matrices in all cases: we integrate the 3 covariant constant 2-forms over a basis of H_2.

The A_n family were previously known due to work of Eguchi-Hanson, Gibbons-Hawking and Hitchin. Also Kronheimer's work has an intimate relation with that of Brieskorn[5] on deformations and resolutions of rational double points.

§4. Yang-Mills moduli spaces

If we accept that hyperkähler 4-manifolds are like algebraic curves then we might conjecture that it should be possible to construct higher dimensional examples as moduli spaces for bundles over "curves". This turns out to be true as I shall now explain.

Let X be a hyperkähler 4-manifold, let G be a compact Lie group and let A be the space of all G-connections for a fixed G-bundle P over X. Then A is an affine space modelled on 1-forms on X with values in g. The I, J, K operators induce similar operators on A which makes A an ∞-dimensional affine space over H, with a compatible metric. Moreover the gauge group $G = \mathrm{Aut}(P)$ acts naturally on A preserving its affine, metric and quaternionic structures. We can therefore consider (rather formally) the ∞-dimensional hyperkähler moment map

$$\mu : A \to (\mathrm{Lie}\ G)^* \otimes R^3$$

and then try to construct hyperkähler quotients.

In fact a little computation (with appropriate care being taken over the non-compactness of X) shows that μ is essentially the self-dual part of the curvature. Thus $\mu = 0$ becomes the <u>(anti)-self-dual Yang-Mills</u> equations which define <u>instantons</u> on X, and the hyperkähler quotient

$$M = \mu^{-1}(0)/G$$

is just the instanton moduli space so extensively studied in general by Donaldson.

There are various cases of special interest, of which the simplest arise for $X = R^4$ or $S^1 \times R^3$. The first gives the instanton moduli spaces studied in [1], while the S^1-invariant part of the moduli space for $S^1 \times R^3$ gives the magnetic monopole moduli spaces studied in [2].

Of course this description is very formal and ignores the analytical difficulties that arise with ∞-dimensional spaces. Nevertheless the analysis works and the conclusions remain valid so that we have here families of hyperkähler manifolds, which arise naturally as <u>hyperkähler quotients of ∞-dimensional affine spaces</u>.

There is actually a very mysterious duality principle of Nahm which means that the same moduli space has 2 different (dual) presentations as a hyperkähler quotient. Roughly speaking Nahm's principle goes as follows. Let $\Gamma \subset R^4$ be a subgroup of the form $R^a \times Z^b$ and let Γ' be the Pontrjagin dual (or character group) of R^4/Γ. Then we can construct Γ-invariant instantons from Γ'-invariant instantons and vice-versa. However, the Lie groups involved are not the same, the duality interchanging quantities like the rank of the Lie group and Chern classes.

In the extreme case when $\Gamma = 0$, then $\Gamma' = R^4$ and Γ'-invariance reduces us to algebra. In fact Nahm's principle, as shown by Donaldson [6], amounts to the main result in [1], [3] which gives an algebraic description of the instanton moduli space. In this case therefore the instanton moduli space has 2 hyperkähler quotient descriptions, one finite-dimensional described in [6] and the other ∞-dimensional.

When $\Gamma = R$ we are in the case originally studied by Nahm and leading to the monopole moduli spaces of [2].

The hyperkähler metric on the instanton moduli spaces of R^4 is, for rather basic reasons, <u>incomplete</u>. For example the first case is $H \times (H-0)/Z_2$ with the flat metric. On the other hand the monopole moduli spaces have <u>complete</u> metrics and this completeness has an important physical interpretation as explained in [2].

These monopole spaces are therefore an interesting class of hyperkähler manifolds and I will return to them in §5. Let me at this stage just say that they have somewhat different asymptotic properties to the ALE spaces.

§5. Twistor Spaces

Twistor spaces were introduced by R. Penrose into theoretical physics with the aim of translating problems from Minkowski space into an alternative framework where complex analysis and geometry can be brought into play. Hyperkähler manifolds fit naturally into the Penrose twistor theory. In fact the 4-dimensional case involving Einstein's equation represents part of the motivation and also the success of the Penrose programme.

The basic idea is very simple. Since a hyperkähler manifold X has a family of complex structures I_λ parametrized by $\lambda \in S^2 = P_1(C)$ we can put all these together on $X \times P_1$. If we put the complex structure I_λ on the fibre $X_\lambda = X \times \{\lambda\}$ and give P_1 its natural complex structure it turns out that we get a complex structure on the total space $Z = X \times P_1$ so that the projection $Z \to P_1$ is a holomorphic map. In terms of the general theory of complex structures we can say that the X_λ form a holomorphic family of complex structures. Z is called the <u>twistor space</u>.

If σ is the antipodal map on $S^2 = P_1$ then $X_{\sigma(\lambda)}$ is the complex conjugate structure to X_λ. Thus $(x,\lambda) \to (x,\sigma(\lambda))$ extends σ to a complex conjugation or real structure on Z. The horizontal sections $\{x\} \times P_1$ are holomorphic curves and are real (i.e. σ-invariant).

By adding a bit more data, essentially the holomorphic symplectic structures on the fibres X_λ we end up with a twistor description (involving only holomorphic data and the real structure σ) which is entirely equivalent to the hyperkähler metric of X.

Returning to our general idea that hyperkähler manifolds provide a substitute for quaternion algebraic varieties the twistor philosophy can now be summarized as follows. Instead of trying to develop a theory of non-commtative quaternionic analysis we use ordinary complex analysis for all embeddings $C \to H$, parametrized by $\lambda \in P_1$, and remember also the holomorphic dependence on λ.

The twistor picture suggests new ways of generating hyperkähler manifolds. For example given a twistor space $Z \to P_1$ we could try to replace each fibre X_λ by a suitable desingularization $X_\lambda(k)$ of its k-fold symmetric product so as to obtain a new twistor space $Z(k) \to P_1$. For this to work the new fibres $X_\lambda(k)$ have to be holomorphic symplectic manifolds. This procedure turns out to work when $\dim X = 4$, so that the X_λ are complex surfaces. The desingularization needed uses Hilbert schemes as in the work of Beauville [4].

Consider in particular the case $X = S^1 \times R^3$ with its standard flat metric. In terms of the magnetic monopoles (for SU(2)) studied in [2] we can identify X with the moduli space M_1 of 1-monopoles: such a monopole has a "location" in R^3 and a "phase" angle. The k-monopole moduli space M_k is, as we observed earlier, a hyperkähler manifold. Its twistor space $Z(k)$ is obtained from the twistor space Z of $M_1 = S^1 \times R^3$ by a version of the desingularized k-fold symmetric product construction indicated above. This means that the horizontal sections of $Z(k) \to P_1$ (which represent points of M_k and hence k-monopoles) correspond to k-sections of $Z \to P_1$, i.e. holomorphic curves meeting each fibre in k points (possibly coincident).

This representation of k-monopoles in R^3 by k-sections of the twistor space is intimately related to "soliton" ideas. I recall that a 1-monopole is viewed as an approximately localized magnetic particle, and a k-soliton can be viewed approximately as a superposition of k such particles <u>provided these are far apart</u>. However when the particles get close together the k-monopole loses its particle identity and is just a complicated non-linear field in space. Translated into the twistor picture this says that a k-monopole, in the far separated case, is represented by a k-section which approximately looks like a union of k simple sections. In general however a k-section does not resemble k separate sections.

The twistor picture enables us to take the soliton idea one stage further. If we fix one fibre of $Z \to P_1$, i.e. if we fix a complex structure of $S^1 \times R^3$, then a k-section does indeed cut this fibre in just k points (possibly coincident), and these determine the k-section. We can in this way think of the general k-monopole as an exact "superposition" of k single monopoles. This description depends however on the choice of complex structure on $S^1 \times R^3$. The dependence is weak in the far separated case (so that we recover the usual soliton picture) but is strong in the nearby (or interactive) case.

Since solitons are one version of the particle/wave dichotomy I like to think that Hamilton, who was much involved in the 19th century controversies on the nature of light, would have been intrigued by the role which quaternions play in connection with solitons.

Let me conclude with a few brief remarks about the first non-trivial monopole space, namely the 2-monopole moduli space M_2. Because there is a natural centre it turns out that, up to a double covering, M_2 is the product of M_1 (representing the centre of mass) and another hyperkähler 4-manifold M_2^o which measures variables relative to the centre. The manifold M_2^o is a very remarkable 4-dimensional hyperkähler manifold and it is extensively studied in [2]. Here are some of its basic properties.

(1) Asymptotically it looks like a circle bundle over $R^3 - 0$,

(2) The fundamental group at ∞ is the quaternion group of order 8,

(3) It admits an action of $SO(3)$ by isometries; this action does not preserve the complex structures, but rotates them,

(4) Its fundamental group is of order 2 and its double covering is the algebraic surface

$$x^2 - zy^2 = 1$$

Property (3) and the hyperkähler property essentially determine the metric uniquely and there is an explicit formula for it involving elliptic integrals. Except for an overall scale there are no free parameters. The geodesics on M_2^o have an interpretation in terms of the dynamics of slowly moving monopoles and this is the main theorem of [2].

I hope this brief and rather sketchy survey of hyperkähler manifolds has shown that they are an interesting class of manifolds which turn up naturally in a variety of places. They bring quaternions into geometry and analysis in a deep and remarkable way and tie up particularly with ideas from theoretical physics. Perhaps they can be viewed as a justification for Hamilton's belief that quaternions have a fundamental role to play in mathematics and physics.

References

1. M.F. Atiyah, Geometry of Yang-Mills Fields, Lezioni Fermiane Accademia Nazionale dei Lincei & Scuola Normale Superiore, Pisa (1979).

2. M.F. Atiyah and N.J. Hitchin, The geometry and dynamics of magnetic monopoles, Princeton University Press (1988).

3. M.F. Atiyah, N.J. Hitchin, V.G. Drinfeld and Y.I. Manin, Construction of instantons, Phys. Letters 65A (1978) 185-7.

4. A. Beauville, Variétés Kähleriennes dont la première classe de Chern est nulle, J. Diff. Geom. 18 (1983), 755-782.

5. E. Brieskorn, Singular elements of semi-simple algebraic groups, Actes Congres Intern. Math. 1970, Vol. 2, 279-284.

6. S.K. Donaldson, Instantons and geometric invariant theory, Commun. Math. Phys. 93 (1984), 453-460.

7. N.J. Hitchin, A. Karlhede, U. Linstrom and M. Roček, Hyperkähler metrics and Supersymmetry, Commun. Math. Phys. 108 (1987), 535-589.

8. F.C. Kirwan, Cohomology of quotients in symplectic and algebraic geometry, Mathematical Notes 31, Princeton Univ. Press (1984).

9. P.B. Kronheimer, Instantons gravitationnels et singularités de Klein, C.R. Acad. Sc. Paris 303 (1986), 53-55.

Affine Differential Geometry and Holomorphic Curves

by Eugenio Calabi

University of Pennsylvania

Consider a smooth, immersed, locally strongly convex surface in euclidean 3-space $\mathbb{R}^3$ oriented so that the second fundamental form II_e is positive definite everywhere, and denote by K_e and dA_e its Gaussian curvature ($K_e > 0$) and element of euclidean area respectively. The following objects are of interest in this note:

(a) the positive density

$$dA = K_e^{1/4}\, dA_e; \tag{0.1}$$

(b) the positive definite quadratic form

$$g = K_e^{-1/4} \mathrm{II}_e; \tag{0.2}$$

(c) the linear functional Y^* on the tangent space of $\mathbb{R}^3$ at each point of the surface, whose value at each vector $\vec{u}$ is

$$Y^*(\vec{u}) = K_e^{-1/4} \langle N_e, \vec{u} \rangle, \tag{0.3}$$

where N_e denotes the unit normal vector to the surface.

The importance of dA, g and Y^* is due to the fact that they are unchanged if one replaces the given euclidean structure on $\mathbb{R}^3$ by any affinely equivalent one inducing the same orientation and volume form; they are in fact the most elementary among the basic *unimodular affine invariants* of the surface. The quadratic form (0.2) is used as an affinely invariant Riemannian metric on the surface, and is called the *Berwald-Blaschke metric;* the positive density (0.1) expresses the corresponding affinely invariant element of area, and the linear functional Y^*, whose null-plane is the tangent plane at each point of the surface, is called the *affine co-normal vector* to the surface.

One of the more obvious questions in affine differential geometry is the one concerning locally strongly convex surfaces that are extremals for the affinely invariant area

$$\int dA = \int K_e^{1/4}\, dA_e \tag{0.4}$$

under interior deformations. The Euler-Lagrange equation for this variational problem is equivalent to the system of equations

$$\Delta Y^* = 0, \tag{0.5}$$

where Δ is the Laplace-Beltrami operator associated to the Berwald-Blaschke metric, applied here independently to each component of the affine co-normal.

The techniques of complex analysis are useful in connection with the Riemannian geometry of surfaces, in the first place because they allow us to simplify many of the otherwise tedious calculations. Accordingly, we shall introduce local complex parameters ζ for the surface, as well as their complex conjugates $\overline{\zeta}$: they are characterized by the orientation of the surface and by the conformal class of the Berwald-Blaschke metric (or, equivalently, of the euclidean second fundamental form II_e). Let the immersion functions of the surface be represented locally as a differentiable mapping X of a parameter domain $\Omega \subset \mathbb{C}$ into $\mathbb{R}^3$ (without loss of generality we shall always assume Ω to be simply connected):

$$X = \begin{pmatrix} x_1(\zeta,\overline{\zeta}) \\ x_2(\zeta,\overline{\zeta}) \\ x_3(\zeta,\overline{\zeta}) \end{pmatrix}, \qquad (\zeta \in \Omega).$$

The relation of the complex structure of the surface, represented by the holomorphy class of the parameter ζ, and the affine invariants (0.1), (0.2), (0.3) is expressed as follows, where (A, B, C) denotes the determinant functional of any ordered triple of vectors A, B, C, and subscripts denote partial derivations with respect to the indicated parameters:

$$\begin{aligned} (X_\zeta, X_{\overline{\zeta}}, X_{\zeta\zeta}) &= 0; \\ (X_\zeta, X_{\overline{\zeta}}, X_{\overline{\zeta}\overline{\zeta}}) &= 0; \\ -i(X_\zeta, X_{\overline{\zeta}}, X_{\zeta\overline{\zeta}}) &> 0. \end{aligned}$$

The Berwald-Blaschke metric then is expressed by

$$ds^2 = 2F\,|d\zeta|^2,$$

where

$$F = \left[-i(X_\zeta, X_{\overline{\zeta}}, X_{\zeta\overline{\zeta}})\right]^{1/2};$$

the affine area element is given by

$$dA = iF\,d\zeta \wedge d\overline{\zeta},$$

and the affine co-normal by

$$Y^* = iF^{-1}\,X_\zeta \times X_{\overline{\zeta}}.$$

The Euler-Lagrange equations (0.5) can be expressed equivalently by the condition that $-\frac{1}{2}Y^*$ is the imaginary part of an immersed holomorphic curve $Z{:}\,\Omega \to \mathbb{C}^3$, uniquely determined by X up to a real translation:

$$Y^*(\zeta,\overline{\zeta}) = i\big(Z(\zeta) - \overline{Z(\zeta)}\big). \tag{0.6}$$

Conversely, Y^* determines X uniquely up to a real translation, as stated more precisely below (see [2]):

Proposition 0.1: *Given any immersed, holomorphic curve $Z:\Omega \to \mathbb{C}^3$, there exists a locally strongly convex surface $X:\Omega \to \mathbb{R}^3$ with $Y^* = i(Z - \overline{Z})$ as its co-normal indicatrix and a Berwald-Blaschke metric conformally related to ζ if and only if Z satisfies the following inequality, in which $Z' = dZ/d\zeta$:*

$$(Z - \overline{Z}, Z', \overline{Z'}) > 0. \tag{0.7}$$

Furthermore, X is uniquely determined by Z up to a translation, and can be expressed explicitely in terms of Z by

$$X = -i\left(Z \times \overline{Z} + \int Z \times dZ - \int \overline{Z} \times \overline{dZ}\right). \tag{0.8}$$

The left-hand member of (0.7) expresses the value of F in terms of Z and ζ.

The inequality (0.7), disregarding a possible change of sign due to a switch of orientation by interchanging the rôles of ζ and $\overline{\zeta}$, has the following geometric interpretation in terms of the holomorphic curve Z: *for each point $Z(\zeta)$ of the holomorphic curve Z, the complex tangent line to it and the line obtained from it by complex conjugation are mutually skew, i.e., not coplanar.*

A locally strongly convex surface X satisfying the Euler-Lagrange equation (0.5), or equivalently (0.6) and (0.8), is called an *affine maximal surface.* Certain important subclasses of affine maximal surfaces have corresponding equivalent descriptions in terms of their representation by holomorphic curves Z as follows:

Proposition 0.2: *A strongly locally convex surface X is a trivial solution of the equation (0.5) characterizing it as an affine maximal surface, i.e., an elliptic paraboloid, if and only if the corresponding holomorphic curve Z describes a complex affine line, necessarily skew to the complex conjugate line.*

Proposition 0.3: *An affine maximal surface X belongs to the special subclass of such surfaces called improper affine spheres, characterized by the property that their co-normal indicatrix traces a plane region, if and only if the corresponding holomorphic curve Z lies in a complex affine 2-plane in $\mathbb{C}^3$ that is parallel to, but distinct from, its complex conjugate.*

The representation of an affine maximal surface in terms of a class of holomorphic curves in $\mathbb{C}^3$ may be interpreted as an affine counterpart of the Weierstrass-Schwartz representation of minimal surfaces in $\mathbb{R}^3$ in terms of "minimal" holomorphic curves in $\mathbb{C}^3$.

The affine analogue of the Plateau problem (which has not been studied as yet) consists of determining an immersion of a closed disk in $\mathbb{R}^3$, with boundary mapped onto a prescribed closed smooth curve, and tangent planes at boundary points in smoothly varying given directions (all compatible with at least one locally strongly convex immersion of the whole disk) in such a way as to *maximize*, whenever possible, the affinely invariant area (0.4) among all the allowable immersions.

For a solution of this problem it will be necessary to find *a priori* estimates for the location of interior point images, as well as of the direction of the tangent plane and of other invariants depending on derivatives of up to third order. One of the conceivable methods that would implicitely furnish estimates for third derivatives in terms of lower order ones consists of solving a "Bernstein problem" for affine maximal surfaces. In its simplest form, this problem amounts to formulating reasonably weak asymptotic conditions at the boundary of the parameter domain, under which any affine maximal surface satisfying them is a trivial solution, i.e., an elliptic paraboloid. The crucial idea in this version of the Bernstein problem exploits the fact that the Gaussian curvature K of the Bernwald-Blaschke metric for an affine maximal surface is non-negative everywhere, and its vanishing identically characterizes the trivial solutions. In fact, for $F(\zeta,\overline{\zeta}) = (Z-\overline{Z}, Z', \overline{Z'})$ according to (0.7), we have

$$K = -F^{-1}\frac{\partial^2 \log F}{\partial\zeta\partial\overline{\zeta}} = F^{-3}A\overline{A},$$

where

$$A = A(\zeta,\overline{\zeta}) = (Z-\overline{Z}, Z', Z''),$$

and $A = 0$ easily implies that $Z' \times Z'' = 0$ or, equivalently, that Z traces holomorphically a complex affine line in $\mathbb{C}^3$.

Theorem 0.4: *Let Z be an immersed holomorphic curve in $\mathbb{C}^3$, parametrized by $\zeta \in \Omega \subset \mathbb{C}$, such that the real valued density*

$$F(\zeta,\overline{\zeta}) = (Z-\overline{Z}, Z', \overline{Z'})$$

is strictly positive for each $\zeta \in \Omega$. If Z satisfies the following two conditions:
(a) *Z is complete with respect to the induced metric*

$$ds^2 = 2F(\zeta,\overline{\zeta})\,|d\zeta|^2; \tag{0.9}$$

(b) *Z lies in a complex affine plane in $\mathbb{C}^3$,*
then Z traces a complex affine line in $\mathbb{C}^3$.

Proof: If the complex curve is complete with respect to the metric (0.9), and since its curvature K is known to be non-negative everywhere, it follows from the theorem of Blanc and Fiala that it is of parabolic type; in particular it implies that the global uniformizing parameter domain Ω is the full complex line $\mathbb{C}$.

If $Z(\mathbb{C})$ lies in a complex affine plane $\Pi \subset \mathbb{C}^3$, it follows that $\Pi \neq \overline{\Pi}$, for if Π coincided with $\overline{\Pi}$, then every tangent line to the curve would be coplanar with its conjugate, and hence $(Z-\overline{Z}, Z', \overline{Z'})$ would be identically zero. We now compactify $\mathbb{C}^3$ to the complex projective 3-space $\mathbb{P}^3(\mathbb{C})$ in the usual way, and denote by Π and $\overline{\Pi}$ the projective completion of the plane containing $Z(\mathbb{C})$ and its conjugate respectively, and consider the complex projective line $\Lambda = \Pi \cap \overline{\Pi} \subset \mathbb{P}^3(\mathbb{C})$. Since $\Lambda = \overline{\Lambda}$, Λ consists of a circle (real projective line) of real points, and two disjoint disks (real hyperbolic planes) H_+ and H_- of imaginary points. Consider the holomorphic map $\varphi\colon \mathbb{C} \to \Lambda$ which assigns to each $\zeta \in \mathbb{C}$ the point of intersection of the tangent line to $Z(\mathbb{C})$ at ζ with Λ. Since $\varphi(\zeta) \neq \varphi(\overline{\zeta})$ (the latter is the intersection of Λ with the conjugate to the tangent line), it follows that the image of $\mathbb{C}$ by φ lies entirely either in H_+ or H_-. It follows from Liouville's theorem that φ is necessarily a constant map, from which one deduces immediately that $Z(\mathbb{C})$ traces a complex affine line, proving the assertion, **q.e.d.**

A corollary of the theorem just proved is the "weak Bernstein theorem" previously announced in [1].

Corollary 0.5: *If X is an immersed, locally strongly convex, affine maximal surface in $\mathbb{R}^3$ which is*
(a) *complete with respect to the Bernwald-Blaschke metric, and*
(b) *globally convex as an (open) surface in $\mathbb{R}^3$,*
then X traces an elliptic paraboloid.

Proof: Since X is an open, globally convex surface in $\mathbb{R}^3$, there exists at least one real affine line L in $\mathbb{R}^3$ such that, for every point of the surface, the line parallel to L at that point intersects there the surface transversally; denoting by $\vec{v}$ a fixed, non-zero vector representing the direction of L, or its opposite, we see that, without loss of generality, $Y^*(\vec{v}) > 0$ everywhere on the surface, where Y^* is the affine co-normal. But the surface is of parabolic type, by the same argument as in Theorem 0.4, and $Y^*(\vec{v})$ is harmonic and everywhere positive: hence $Y^*(\vec{v})$ is constant. Since $i(Z - \overline{Z}) \cdot \vec{v}$ is constant, it follows that $Z \cdot \vec{v}$ is constant, so that Z lies in a complex affine plane, and the argument follows as in Theorem 0.4, **q.e.d.**

The unsatisfactory feature of the statement of this corollary as a "Bernstein theorem" is that it is hard to conceive of the two completeness assumptions, the metric one in terms of the Berwald-Blaschke metric and the affine one in terms of global convexity of the surface, as being both independently essential, especially since neither of the two assumptions seems to be exploited to its fullest extent. Howewer, we have not succeded so far in deducing either one of these two assumptions as a consequence of the other one. On the other hand we can prove another somewhat sharper version of the corollary, resulting by substituting another sort of a restrictive assumption, which still seems unnecessarily strong, as follows:

Theorem 0.6: *Let Z be an entire, immersed holomorphic curve, i.e., $Z\colon \mathbb{C} \to \mathbb{C}^3$ with $Z'(\zeta) \neq 0$ everywhere. Assume that, for any two points ζ_1, $\zeta_2 \in \mathbb{C}$ the complex tangent line to the curve at ζ is skew with respect to the complex conjugate to the tangent line at ζ_2, i.e.,*

$$(Z(\zeta_1) - \overline{Z(\zeta_2)}, Z'(\zeta_1), \overline{Z'(\zeta_2)}) \neq 0. \tag{0.10}$$

Then $Z(\mathbb{C})$ lies in a complex affine line.

Proof: We first compose the immersion Z with the natural imbedding of $\mathbb{C}^3$ in $\mathbb{P}^3(\mathbb{C})$ by adjoining the constant 1 before the three components of Z, and treating the four resulting functions as homogeneous coordinates representing an immersion in $\mathbb{P}^3(\mathbb{C})$, denoted by Z_p. Then the six homogeneous Plücker coordinates of the tangent lines to the curve are represented, up to a scalar factor, by $Z_p \wedge Z'_p$ and can be described equivalently by the six components of the pair of vector functions $(Z', Z \times Z')$.

Lemma 0.7: *The linear span in $\mathbb{C}^6$ of the values of the six Plücker coordinates $Z_p \wedge Z'_p$ of the tangent lines of $Z(\mathbb{C})$ is at most three-dimensional over $\mathbb{C}$.*

Proof of the Lemma: Denote by k the complex dimension of the linear span of $Z_p \wedge Z'_p$ (a priori, $1 \leq k \leq 6$). Consider the inequality (0.10), at first for a fixed ζ_2, as ζ_1 varies over $\mathbb{C}$. This can be rewritten as the (4×4)-determinant equation

$$Z_p(\zeta_1) \wedge Z'_p(\zeta_1) \wedge \overline{Z_p(\zeta_2)} \wedge \overline{Z'_p(\zeta_2)} \neq 0;$$

for fixed ζ_2 this implies one homogeneous linear constraint on $Z_p \wedge Z'_p \colon \mathbb{C} \to \mathbb{C}^6$; in other words, the image of $\mathbb{C}$ in $\mathbb{P}^5(\mathbb{C})$ represented by treating $Z_p \wedge Z'_p$ as homogeneous coordinates is disjoint from the hyperplane dual to the point $\overline{Z_p \wedge Z'_p(\zeta_2)}$.

We now repeat the argument taking all values of ζ_2, at least in a set that is dense in some open subdomain of $\mathbb{C}$. This shows that the image of $\mathbb{C}$ in $\mathbb{P}^5(\mathbb{C})$ under $\zeta_1 \mapsto (Z_p \wedge Z'_p)(\zeta_1)$ avoids an infinite collection of hyperplanes, whose common intersection is a projective subspace of $\mathbb{P}^5(\mathbb{C})$, exactly $(5-k)$-dimensional. It follows from the Jensen-Nevanlinna theory, or even the most elementary generalization of Picard's theorem to higher dimensions, that $(Z_p \wedge Z'_p)(\mathbb{C})$ is contained in the join of one of its point with that $(5-k)$-dimensional projective subspace. But that number again is *at least* equal to $k-1$. This implies that $2k \leq 7$, and so it follows that $k \leq 3$, as claimed, **q.e.d.**

To complete the proof of Theorem 0.6, we observe that the image of $\mathbb{C}$ in $\mathbb{P}^5(\mathbb{C})$ represented by $Z_p \wedge Z'_p$ or equivalently by $Z' \oplus (Z \times Z')$ lies in the intersection of a complex projective 2-subplane with the nonsingular quadric described by the identity $Z' \cdot (Z \times Z') = 0$. Such an intersection is either:

(a) a projective 2-plane in the Grassmannian of all projective lines in $\mathbb{P}^3(\mathbb{C})$ (the quadric in $\mathbb{P}^5(\mathbb{C})$ is just that), consisting of all lines through a fixed point in $\mathbb{P}^3(\mathbb{C})$, and this is clearly absurd;

(b) a projective 2-plane in the same Grassmannian consisting of all lines in a 2-subplane of $\mathbb{P}^3(\mathbb{C})$; this would reduce $Z(\mathbb{C})$ to a plane curve in $\mathbb{C}^3$, where the conclusion follows by Theorem 0.4;

(c) a reducible quadric curve if the 2-plane is tangent to but not contained in the 4-dimensional quadric; in this case, since the image of $\mathbb{C}$ is irreducible, it would be a line in $\mathbb{P}^5(\mathbb{C})$, and this would again lead to an absurdity;

(d) finally, it could be a generic intersection. This would mean that the image of $\mathbb{C}$ in $\mathbb{P}^5(\mathbb{C})$ is relatively open in an irreducible conic section: the family of lines in $\mathbb{P}^3(\mathbb{C})$ represented by this curve in the Grassmannian quadric in $\mathbb{P}^5(\mathbb{C})$ consists of the rulings on an irreducible quadric surface in $\mathbb{P}^3(\mathbb{C})$, either singular or nonsingular. Since neither of these two cases can occur as the family of tangents to an immersed holomorphic curve, the theorem is completely proved, **q.e.d.**

A subsequent study on this topic should lead to a *quantitative* counterpart of the "Bernstein theorem" just established, or possibly of a sharper version of it. This means to seek a result of the following type. Consider a locally strongly convex, affine maximal surface in $\mathbb{R}^3$ that is not trivial, as measured quantitatively, for instance, by a positive value K_0 of the curvature of the Berwald-Blaschke metric at some interior point p_0; one would want to estimate an upper bound for K_0 in terms of the distance, defined geodesically or otherwise, of p_0 from the nearest singularity or "ideal boundary point"; the latter measure is to be interpreted as a quantitative expression of the extent to which the conditions for the Bernstein theorem are not satisfied.

References.

[1] E. Calabi: *Hypersurfaces with maximal affinely invariant area.* Amer. J. Math. **104** (1982), 91–126.

[2] E. Calabi: *On affine maximal surfaces.* Technical report, Technische Universität, Berlin, 1987 (to appear).

The Meromorphic Continuation of Kloosterman-Selberg Zeta Functions

J.W. COGDELL[1] AND I.I. PIATETSKI-SHAPIRO[2]

Introduction.

Let $S(n)$ denote the classical Kloosterman sum

$$S(n) = \sum_{\substack{x,y (\text{mod } n) \\ xy \equiv 1 (\text{mod } n)}} e^{2\pi i (\frac{x+y}{n})}.$$

The Kloosterman-Selberg zeta function of level N is then

$$Z_N(s) = \sum_{\substack{n=0 \\ n \equiv 0 (\text{mod } N)}}^{\infty} \frac{s(n)}{n^{2s}}.$$

From the trivial estimate $|S(n)| \leq n$ on the Kloosterman sums one sees that $Z_N(s)$ defines an analytic function for $Re(s) > 1$. In his celebrated paper of 1965 on the estimation of Fourier coefficients of modular forms Selberg studied the analytic properties of $Z_N(s)$ and related functions [**5**]. He showed that $Z_N(s)$ has a meromorphic continuation to all of $\mathbb{C}$ and that the location of the poles are related to the arithmetic of the congruence subgroups

$$\Gamma_0(N) = \left\{ \begin{pmatrix} a & b \\ c & d \end{pmatrix} \in SL_2(\mathbb{Z}) \mid c \equiv 0 \ (\text{mod } N) \right\}.$$

In particular, he showed that the poles of $Z_N(s)$ in $Re(s) \geq \frac{1}{2}$ are in a one-to-one correspondence with the discrete eigenfunctions of the Laplacian Δ on $\Gamma_0(N)\backslash\mathfrak{H}$ having non-vanishing first Fourier coefficient. If φ is such a discrete eigenfunction with $\Delta\varphi = \lambda\varphi$ then, writing $\lambda = \frac{1}{4} - \rho^2$, $Z_N(s)$ will have a pole at $s = \frac{1}{2} + \rho$. Selberg did not restrict himself to $\Gamma_0(N)$, but dealt with a larger class of Fuchsian groups. The above is indicative of the type of results he discussed.

---

[1]Department of Mathematics, Oklahoma State University, Stillwater, OK 74078
Partially supported by an NSF grant.
[2]Department of Mathematics, Yale University, New Haven, CT 06520 and School of Mathematics, The Raymond and Beverly Sackler Faculty of Exact Sciences, Tel-Aviv University, Tel-Aviv, Israel
Partially supported by an NSF grant.

In order to prove this result, Selberg constructed Poincaré series depending o a complex parameter s. He found a beautiful proof of the meromorphic continu ation of these Poincaré series using a shift equation. He showed that the Fourie coefficients of these Poincaré series had the form

$$c_{\psi}(s) = \sum_{n} \lambda_n(s) Z_N(s+n)$$

and from this formula he deduced the meromorphic continuation of $Z_N(s)$. Thi type of method is difficult to generalize to higher dimensions.

In this paper we would like to indicate generalizations of this type result in tw directions. First we consider the Kloosterman-Selberg zeta function $Z_\Gamma(s)$ for dis crete subgroups Γ acting on Lobaehevskii space of arbitrary dimension, i.e., discret subgroups Γ of $SO(n,1)$. For these zeta functions we will give a meromorphic con tinuation and locate the poles. We also consider arbitrary Fuchsian groups Γ of th first kind in $SL_2(\mathbb{R})$, and for these we will derive a precise formula for $Z_\Gamma(s)$ fron which the continuation follows. Note that in both cases, as in Selberg's origina results, it is not merely the meromorphic continuation of $Z_\Gamma(s)$ which is of interes but the actual location of the poles of this continuation.

The method which we use allows us to give a direct integral representation o $Z_\Gamma(s)$ as the Mellin-Whittaker transform of some smooth function $P(g)$ on $\Gamma\backslash G$ The analytic continuation of this Mellin-Whittaker transform is more complicated One of the tools which we use is a trick suggested by J. Bernstein which present a generalization of Selberg's shift argument. Our basic approach is based on "soft Kuznetsov formula" which we can formulate for discrete subgroups Γ of an semi-simple Lie group G over $\mathbb{R}$ such that the quotient $\Gamma\backslash G$ is not compact bu still has finite volume. We expect that these techniques will eventually lead to th meromorphic continuation of the Kloosterman-Selberg zeta functions in this genera setting.

We would like to thank J. Bernstein for many helpful discussions.

1. The Soft Kuznetsov Formula.

Let G be any semi-simple Lie group over $\mathbb{R}$ and let $\Gamma \subset G$ be a discrete subgrou such that $\Gamma\backslash G$ is non-compact but has finite volume. Let $P \subset G$ be a paraboli subgroup, $P = MU$ with M its Levi component and U its unipotent radical. Let M^0 be the intersection of the kernels of $|\chi|$ for all continuous characters $\chi : M \to \mathbb{R}^\times$ We say that P is Γ-rational if $(\Gamma \cap U)\backslash U$ is compact. In this case if we let Γ_{M^0} b the projection of $\Gamma \cap P^0$ onto M^0, then $\Gamma_{M^0}\backslash M^0$ again has finite volume.

A smooth function φ on $\Gamma\backslash G$ is called cuspidal if for every Γ-rational paraboli $P = MU$ we have

$$\int_{(\Gamma\cap U)\backslash U} \varphi(ug)du = 0$$

or all $g \in G$.

Consider the space $L^2(\Gamma\backslash G)$. This gives a unitary representation of G acting by right translation. $L^2(\Gamma\backslash G)$ may then be decomposed into irreducible unitary representations of G. As a preliminary decomposition we have

$$L^2(\Gamma\backslash G) = L_0^2(\Gamma\backslash G) \oplus L_{\mathbf{res}}^2(\Gamma\backslash G) \oplus L_{\mathbf{cont}}^2(\Gamma\backslash G).$$

$L_0^2(\Gamma\backslash G)$ is the invariant subspace spanned by all cuspidal functions on $\Gamma\backslash G$. It further decomposes into a discrete direct sum of irreducible representations, each isomorphism class occurring with finite multiplicity. $L_{\mathbf{res}}^2(\Gamma\backslash G)$ is the invariant subspace spanned by residues of Eisenstein series. It decompsoes into a discrete sum of irreducible representations. Finally, $L_{\mathbf{cont}}^2(\Gamma\backslash G)$ is a direct sum of continuous direct sums of irreducible representations built from Eisenstein series induced off of equivalence classes of the Γ-rational parabolics. Set $L_{\mathbf{dis}}^2(\Gamma\backslash G) = L_0^2(\Gamma\backslash G) \oplus L_{\mathbf{res}}^2(\Gamma\backslash G)$. Then we will write this decomposition as

$$L^2(\Gamma\backslash G) = (\bigoplus_{\pi \subset L_{\mathbf{dis}}^2(\Gamma\backslash G)} \pi) \oplus \sum_{\{P\}} \sum_{\tau \subset L_0^2(\Gamma_M^0\backslash M^0)} \int \pi(\tau, r)\, d\mu(r)$$

where $\pi(\tau, r)$ is an induced representation from P to G.

Now fix N the unipotent subgroup of some Γ-rational parabolic. Let $\Delta = \Gamma \cap N$, so that $\Delta\backslash N$ is compact. Let ψ be a non-trivial character of N which is trivial on Δ. Let $S(N\backslash G, \psi)$ denote the space of smooth functions f on G satisfying

$$f(ng) = \psi(n) f(g) \qquad n \in N.$$

Then for $f \in S(N\backslash G, \psi)$ we may form the Poincaré series

$$P_f(g) = \sum_{\gamma \in \Delta\backslash\Gamma} f(\gamma g).$$

Under suitable regularity conditions on f (for example, compact support mod N) $P_f(g)$ converges to a smooth function in $L^2(\Gamma\backslash G)$. Then $P_f(g)$ will have a spectral decomposition according to the decomposition of $L^2(\Gamma\backslash G)$ above which we will now describe.

For irreducible π occurring discretely in $L^2(\Gamma\backslash G)$ and smooth $\varphi \in \pi$, let

$$W_\varphi(g) = \int_{\Delta\backslash N} \varphi(ng)\psi^{-1}(n)\, dn$$

be the associated Whittaker function. Then f will define a functional on π via

$$\varphi \mapsto \int_{N\backslash G} f(g)\overline{W_\varphi(g)}\, dg.$$

Then there will exist a smooth vector $F_\pi(f) \in \pi$ representing this functional, i.e such that

$$(F_\pi(f), \varphi)_{\Gamma\backslash G} = \int\limits_{N\backslash G} f(g)\overline{W_\varphi(g)}dg$$

where the inner product on the left is the natural Petersson inner product o $L^2(\Gamma\backslash G)$. Similarly, one can introduce the projections $F_{\pi(\tau,r)}(f)$ to the irreducib representations $\pi(\tau, r)$ occurring in the continuous spectrum. The Poincaré serie then has a decomposition

$$P_f = \sum_{\pi \subset L^2_{\mathrm{disc}}(\Gamma\backslash G)} F_\pi(f) + \sum_{\{P\}} \sum_{\tau \subset L^2_0(\Gamma_{M^0}\backslash M^0)} \int F_{\pi(\tau,r)}(f) d\mu(r).$$

Let us now introduce the Kloosterman sums relevant to this situation. With I and ψ fixed as above, let

$$\Omega(\Gamma) = \{g \in G \mid \Gamma \cap NgN \neq \phi\}.$$

Then for $g \in \Omega(\Gamma)$ and $\gamma \in NgN \cap \Gamma$ we may write $\gamma = n_1(\gamma) g n_2(\gamma)$ with $n_i(\gamma) \in N$ The ψ-Kloosterman sum associated to $g \in \Omega(\Gamma)$ is then

$$Kl_\psi(g) = \sum_{\gamma \in \Delta\backslash\Gamma\cap NgN/\Delta} \psi(n_1(\gamma))\psi(n_2(\gamma)).$$

Furthermore, for $g \in \Omega(\Gamma)$ and $f \in S(N\backslash G, \psi)$ let us define

$$M(g, f) = \int\limits_{N_g\backslash N} f(gn)\psi^{-1}(n)dn$$

where N_g is the stabilizer of g in N under conjugation.

If φ is any smooth function in $L^2(\Gamma\backslash G)$ then we may compute its ψ-Fourie coefficient, by which we mean

$$C_\psi(\varphi) = W_\varphi(1) = \int\limits_{\Delta\backslash N} \varphi(n)\psi^{-1}(n)dn.$$

If we do this for our Poincaré series $P_f(g)$ we may compute in two ways. First, we unfold the Poincaré series we arrive at

$$C_\psi(P_f) = \sum_{g \in N\backslash\Omega(g)/N} Kl_\psi(g) M(g, f).$$

On the other hand, from the spectral expansion we have

$$C_\psi(P_f) = \sum_{\pi\subset L^2_{\text{disc}}(\Gamma\backslash G)} C_\psi(F_\pi(f)) + \sum_{\{P\}}\sum_{\tau}\int C_\psi(F_{\pi(\tau,r)}(f))d\mu(r).$$

Equating these two expressions gives the soft Kuznetsov formula, namely,

$$(1.1)\qquad \sum_{g\in N\backslash\Omega(\Gamma)/N} Kl_\psi(g)M(g,f) = \sum_{\pi\subset L^2_{\text{disc}}(\Gamma\backslash G)} C_\psi(F_\pi(f)) + \sum_{\{P\}}\sum_{\tau}\int C_\psi(F_{\pi(\tau,r)}(f))d\mu(r).$$

2. The Kloosterman-Selberg Zeta Functions for Lobachevskii Space.

Now let $G = SO(r+1,1)$. The symmetric space of G is then $r+1$ dimensional Lobachevskii space

$$L_{r+1} = \{\vec{x}\in\mathbb{R}^{r+1} \mid x_1^2+\cdots+x_{r+1}^2 < 1\}.$$

G will act on this space via linear fractional transformations. The boundary of this space is B_r, the unit sphere in $\mathbb{R}^{r+1}$. If $\vec{x}\in B_r$ is on the boundary of Lobachevskii space, then the stabilizer of $\vec{x}$ in G will be a minimal parabolic subgroup $P = MU$ with abelian unipotent radical $U\simeq\mathbb{R}^r$ and Levi component $M\simeq\mathbb{R}^\times\times SO(r)$.

Let $\Gamma\subset G$ be a discrete subgroup of the type considered in Section 1. Then Γ gives a discrete group of motions on Lobachevskii space. A point $\vec{x}\in B_r$ on the boundary of Lobachevskii space is called a cusp of Γ if the associated parabolic subgroup P is Γ-rational.

Fix now a cusp for Γ and write the associated parabolic as $P = MN$ with N the unipotent radical. By the Bruhat decomposition for G we have $G = P\cup NwP$ with w the non-trivial element of the Weyl group of G. Hence we have canonical representatives for the double cosets $N\backslash G/N$ via $N\backslash G/N\simeq M\cup wM$. Set now

$$M(\Gamma) = \{m\in M \mid \Gamma\cap NwmN\neq\phi\}.$$

Letting $\Delta = \Gamma\cap N$ as before we see that we may take representatives for $N\backslash\Omega(\Gamma)/N$ of the form wm with $m\in M(\Gamma)$. Since $M\simeq\mathbb{R}^\times\times SO(r)$, we may write $m = a_m h_m$ with $a_m\in\mathbb{R}^\times$ and $h_m\in SO(r)$.

Fix now a non-trivial character ψ of $\Delta\backslash N$. Also fix a finite dimensional representation τ of $H = SO(r)$ and let $\xi_\tau(h)$ be a matrix coefficient of τ. Then the relevant Kloosterman-Selberg zeta function for this situation is

$$Z_\Gamma(\tau,s) = \sum_{m\in M(\Gamma)} Kl_\psi(wm)\xi_\tau(h_m)|a_m|^{-(s+\frac{r}{2})}.$$

By comparing with the appropriate Eisenstein series to get the trivial bound on Kloosterman sums, it follows that the series for $Z_\Gamma(\tau, s)$ converges absolutely for $Re(s) > \frac{r}{2}$.

To analyze this function using the soft Kuznetsov formula, we need an appropriate Poincaré series or equivalently an appropriate $f \in S(N\backslash G, \psi)$. To construct f we recall the Bruhat decomposition $G = P \cup NwP$. Let $\mu \in C_c^\infty(\mathbb{R}^\times)$ and $\nu \in \mathcal{S}(\mathbb{R}^r)$. Since $N \simeq \mathbb{R}^r$ we may consider ν as an element of $\mathcal{S}(N)$. Then on the big Bruhat cell set

$$f(n_1 w n_2 m) = \psi(n_1)\nu(n_2)\mu(a_m)\xi_\tau(h_m)$$

and on the small cell set $f(p) = 0$ for $p \in P$. Now, if we let $y \in \mathbb{R}^\times \subset M$, then $R(y)f \in S(N\backslash G, \psi)$ where $R(y)$ denotes right translation by y. The associated Poincaré series $P_{R(y)f}(g)$ is in $L^2(\Gamma\backslash G)$ and the soft Kuznetsov formula gives

(2.1)
$$\sum_{m \in M(\Gamma)} Kl_\psi(wm) M(wm, R(y)f) = \sum_{\pi \subset L^2_{\text{disc}}(\Gamma\backslash G)} C_\psi(F_\pi(R(y)f)) + \sum_{\{P\}} \sum_{\sigma \in \hat{H}} \int C_\psi(F_{\pi(\sigma,r)}(R(y)f)) d\mu(r).$$

By an application of Sobolev's lemma we can show this absolutely converges pointwise in y. Furthermore, we may compute the Mellin transform term by term. If for any smooth function $\varphi \in L^2(\Gamma\backslash G)$ we set

$$I_\varphi(\tau, s) = \int_M W_\varphi(m)\overline{\xi_\tau(h_m)}|a_m|^{s-\frac{r}{2}} dh_m d^\times a_m$$

Then our soft Kuznetsov yields

(2.2)
$$\sum_{m \in M(\Gamma)} Kl_\psi(wm) \int_0^\infty M(wm, R(y)f) y^{s-\frac{r}{2}} d^\times y$$
$$= \sum_{\pi \subset L^2_{\text{disc}}(\Gamma\backslash G)} I_{F_\pi(f)}(\tau, s) + \sum_{\{P\}} \sum_{\sigma \in \hat{H}} \int I_{F_{\pi(\sigma,r)}(f)}(\tau, s) d\mu(r)$$

absolutely convergent for $Re(s) > \frac{r}{2} + 1$. Note that the left hand side may be interpreted as a computation of $I_{P_f}(\tau, s)$.

Suppose that our character ψ is given as follows. As $N \simeq \mathbb{R}^r$, for $n \in N$ let $\vec{n}$ denote the associated vector in $\mathbb{R}^r$. Let $\vec{\alpha} \in \mathbb{R}^r$ be such that $\psi(n) = e^{2\pi i \vec{\alpha} \cdot \vec{n}}$. For

$\nu \in \mathcal{S}(\mathbb{R}^r)$, let $\hat{\nu}$ denote the Fourier transform with respect to ψ^{-1}. Then a simple computation gives

$$\int_0^\infty M(wm, R(y)f)y^{s-\frac{r}{2}}d^\times y$$

$$= \hat{\nu}\left(\frac{1}{a_m}h_m\vec{\alpha}\right)\tilde{\mu}\left(s-\frac{r}{2}\right)|a_m|^{-(s+\frac{r}{2})}\xi_\tau(h_m),$$

where for $\mu \in C_c^\infty(\mathbb{R}^\times)$ we let $\tilde{\mu}(s) = \int_0^\infty \mu(y)y^s d^\times y$. By the discreteness of Γ, there exists a constant C such that for $m \in M(\Gamma), |a_m| > C$. Since $h_m \in SO(r)$, a compact group, we see that $\{\frac{1}{a_m}h_m\vec{\alpha} \mid m \in M(\Gamma)\}$ is contained in a compact subset of $\mathbb{R}^r$. Hence there exists $\nu \in \mathcal{S}(\mathbb{R}^r)$ such that $\hat{\nu}$ is identically one on this set. If we make this choice of ν, then the left hand side of (2.2) becomes $\tilde{\mu}(s-\frac{r}{2})Z_\Gamma(\tau, s)$. Hence we have

PROPOSITION 1. *For an appropriate choice of $\nu \in \mathcal{S}(\mathbb{R}^r)$ and any $\mu \in C_c^\infty(\mathbb{R}^\times)$ we have*

$$\tilde{\mu}(s-\frac{r}{2})Z_\Gamma(\tau, s) = \sum_{\pi \subset L^2_{\mathrm{disc}}(\Gamma\backslash G)} I_{F_\pi(f)}(\tau, s)$$

$$+\sum_{\{P\}}\sum_{\sigma\in\hat{H}}\int I_{F_{\pi(\sigma,r)}(f)}(\tau, s)d\mu(r)$$

for $Re(s) > \frac{r}{2}+1$, where f is defined in terms of μ and ν as above.

As we noted above, the left hand side of Proposition 1 may be interpreted as

$$I_{P_f}(\tau, s) = \tilde{\mu}(s-\frac{r}{2})Z_\Gamma(\tau, s)$$

for an appropriate choice of f. Hence we have reduced the meromorphic continuation of $Z_\Gamma(\tau, s)$ to the seemingly more general problem of meromorphically continuing $I_\varphi(\tau, s)$ for φ any smooth function in $L^2(\Gamma\backslash G)$.

The analytic continuation of $I_\varphi(\tau, s)$ for smooth φ is based on the following two lemmas.

LEMMA 1. *Let $\varphi \in \pi \subset Ind_P^G(|\ |^\rho \otimes \sigma)$ be smooth. ($|\ |^\rho \otimes \sigma$ is a representation of P via $mn \mapsto |a_m|^\rho\sigma(h_m)$). Then $I_\varphi(\tau, s)$ has a meromorphic continuation to all of $\mathbb{C}$ with possible poles at $s = \pm\rho - k$ for $k = 0, 1, 2, \cdots$. If $\rho \neq 0$ these poles are simple. $I_\varphi(\tau, s)$ can have a pole at $\pm\rho - k$ only if τ is a constituent of $\sigma \otimes R^k$ where R is the standard representation of $H = SO(r)$ on $\mathbb{R}^r$.*

The point of this lemma is that if φ lies in an irreducible representation π, then formulas of Wallach [6] and Casselman [1] give explicit asymptotics for $W_\varphi(m)$ as $a_m \to 0$. From these asymptotics, the continuation of $I_\varphi(\tau, s)$ is routine.

The second lemma is a shift equation due to J. Bernstein. We use such a shift equation for each generator of the center of the universal enveloping algebra $U(\mathfrak{g})$ of $\mathfrak{g}$, the Lie algebra of G. For the Casimir operator this shift equation takes the following form.

LEMMA 2. *With appropriate normlizations, let D be the Casimir operator of G, D_H the Casimir of H. For π an irreducible representation of G and τ an irreducible representation of H let $\pi(D) = \lambda_\pi$ and $\tau(D_H) = \lambda_\tau$ be the Casimir eigenvalues. Let $\varphi \in \pi$ be smooth. Then*

$$I_\varphi(\tau, s) = \frac{-4\pi i}{s^2 + rs + \lambda_\pi + 2(r-2)\lambda_\tau} \sum_{i=1}^{r} I_{Y_i\varphi}(\tau \otimes R, s+1)$$

where Y_i are a basis for the Lie algebra of N^- the opposite unipotent subgroup and $I_{Y_i\varphi}(\tau \otimes R, s+1)$ is defined with an appropriate matrix element of $\tau \otimes R$.

This lemma is proven by computing $I_{D\varphi}(\tau, s)$ in two ways, once as $\lambda_\pi I_\varphi(\tau, s)$ and the other by explicitly writing $D = \Sigma X_i X_i^\wedge$ for $\{X_i\}$ an appropriate basis of $\mathfrak{g}$ and computing the action term by term. The other generators of the center of $U(\mathfrak{g})$ are treated similarly. Again, we use heavily that φ actually lies in an irreducible representation.

Now let us again look at φ an arbitrary smooth function in $L^2(\Gamma \backslash G)$. Then φ will have a spectral decomposition

$$\varphi = \sum_{\pi \subset L^2_{\text{disc}}(\Gamma\backslash G)} F_\pi(\varphi) + \sum_{\{P\}} \sum_{\sigma \in \hat{H}} \int F_{\pi(\sigma,r)}(\varphi) d\mu(r)$$

and an application of Sobolev's lemma gives that for $Re(s)$ sufficiently large

$$I_\varphi(\tau, s) = \sum_{\pi} I_{F_\pi(\varphi)}(\tau, s) + \sum_{\{P\}} \sum_{\sigma} \int I_{F_{\pi(\sigma,r)}(\varphi)}(\tau, s) d\mu(r).$$

For $\varphi = P_f$, this is exactly Proposition 1. Let

$$I^d_\varphi(\tau, s) = \sum_{\pi \subset L^2_{\text{disc}}(\Gamma\backslash G)} I_{F_\pi(\varphi)}(\tau, s)$$

$$I^c_\varphi(\tau, s) = \sum_{\{P\}} \sum_{\sigma} \int I_{F_{\pi(\sigma,r)}(\varphi)}(\tau, s) d\mu(r).$$

Then to continue $I_\varphi(\tau, s)$ it is sufficient to continue $I^d_\varphi(\tau, s)$ and $I^c_\varphi(\tau, s)$ seperately. For now, we will restrict our attention to $I^d_\varphi(\tau, s)$. Using the above Lemmas, we have:

PROPOSITION 2. *$I_\varphi^d(\tau,s)$ has a meromorphic continuation to all of $\mathbb{C}$. It has possible double poles at $s=-k$ for $k=0,1,\cdots$ and at most single poles for $s=\pm\rho-k$ which can only occur if there exists $\pi\subset L^2_{\mathrm{disc}}(\Gamma\backslash G)$ such that $\pi\subset \mathrm{Ind}(|\ |^\rho\otimes\sigma)$ and τ is a constituent of $\sigma\otimes R^k$.*

The continuation of $I_\varphi^c(\tau,s)$ is achieved by a similar method and gives possible poles at the poles of the Eisenstein series for Γ. If we now return to our original problem, combining the above with Proposition 1 gives the following result.

THEOREM 1. *The Kloosterman-Selberg zeta function $Z_\Gamma(\tau,s)$ has a meromorphic continuation to all of $\mathbb{C}$. It has possible poles for $Re(s)\le\frac{r}{2}$ located at*

(i) the points $s=\pm\rho-k$ if there exists a $\pi\subset L^2_{\mathrm{disc}}(\Gamma\backslash G)$ such that $\pi\subset \mathrm{Ind}(|\ |^\rho\otimes\sigma)$ with τ a constituent of $\sigma\otimes R^k$

(ii) the points $s=0,-1,-2,\cdots$

(iii) the poles of the Eisenstein series for Γ.

The complete details of the continuation as well as other results for Lobachevskii space are to be found in [**2**].

3. The Kloosterman-Selberg Zeta Function for Fuchsian Groups of the First Kind.

Now let $G=SL_2(\mathbb{R})$ and take $\Gamma\subset G$ to be an arbitrary Fuchsian group of the first kind. Let $B=\left\{\begin{pmatrix} a & b\\ 0 & a^{-1}\end{pmatrix}\right\}$ be the Borel subgroup of G and $N=\left\{\begin{pmatrix} 1 & x\\ 0 & 1\end{pmatrix}\right\}$ its unipotent radical. For simplicity let us assume that $\Gamma\cap N=\Delta=\left\{\begin{pmatrix} 1 & n\\ 0 & 1\end{pmatrix}\ \middle|\ n\in\mathbb{Z}\right\}$. This can always be arranged by conjugation within $SL_2(\mathbb{R})$. Also take $\psi\left(\begin{pmatrix} 1 & x\\ 0 & 1\end{pmatrix}\right)=e^{2\pi i x}$ for our additive character.

Before considering the Kloosterman-Selberg zeta function, let us show how our soft Kuznetsov formula gives a version of the Kuznetsov trace formula. We will then use this formula to derive a precise formula for the Kloosterman-Selberg zeta function. In the soft Kuznetsov formula, we form our Poincaré series P_f out of a function $f(g)$ defined using the Bruhat decomposition $G=B\cup NwB$ where $w=\begin{pmatrix} 0 & -1\\ 1 & 0\end{pmatrix}$. Take $\mu\in C_c^\infty(\mathbb{R}^\times)$ and $\nu\in\mathcal{S}(\mathbb{R})$ and set

$$f(g)=\begin{cases}\psi(x_1)\nu(x_2)\mu(y) & g=\begin{pmatrix} 1 & x_1\\ 0 & 1\end{pmatrix} w\begin{pmatrix} 1 & x_2\\ 0 & 1\end{pmatrix}\begin{pmatrix} y & 0\\ 0 & y^{-1}\end{pmatrix}\\ 0 & g\in B\end{cases}$$

Hence $f(g)$ is essentially supported in a neighborhood of w on the big Bruhat cell.

If we substitute this function into (1.1) then the left hand side is computable as in Section 2. If we now set

$$M(\Gamma) = \left\{c \in \mathbb{R}^\times \mid Nw \begin{pmatrix} c & 0 \\ 0 & c^{-1} \end{pmatrix} N \cap \Gamma \neq \phi\right\}$$

and let $A(x) = \hat{\nu}(x^2)\mu(\frac{1}{x})$ then the left hand side of (1.1) becomes

$$\sum_{c \in M(\Gamma)} \frac{Kl_\psi(c)}{|c|^2} A(\frac{1}{c})$$

where we have set $Kl_\psi(c) = Kl_\psi(w \begin{pmatrix} c & 0 \\ 0 & c^{-1} \end{pmatrix})$.

On the other hand, for $\pi \in L^2(\Gamma \backslash G)$ occurring discretely we may now explicitly compute the Fourier coefficients of the projections $F_\pi(f)$ from the equality

$$(F_\pi(f), \varphi)_{\Gamma \backslash G} = (P_f, \varphi)_{\Gamma \backslash G} \quad \varphi \in \pi.$$

We compute $(P_f, \varphi)_{\Gamma \backslash G}$ in terms of the Kirillov model of π on $L^2(\mathbb{R}^\times)$ associated with the character ψ [**3**]. Since our f is essentially supported near w, the computation will involve the Bessel function of π, which we denote by $J_\pi(y)$, which gives the action of w in the Kirillov model of π. What we finally find is that, again letting $A(x) = \hat{\nu}(x^2)\mu(\frac{1}{x})$ as above,

$$C_\psi(F_\pi(f)) = c(\pi, \Gamma) \int_{\mathbb{R}^\times} A(x) J_\pi(x) d^\times x$$

where $c(\pi, \Gamma)$ is an arithmetic constant of proportionality between the two G-invariant inner products on π, namely the Petersson inner product from $L^2(\Gamma \backslash G)$ and the Kirillov inner product from the model on $L^2(\mathbb{R}^\times)$. The constants $c(\pi, \Gamma)$ are explicitly computable in terms of classical Fourier coefficients and the parameter of the representation. A similar computation is valid for the continuous spectrum of $L^2(\Gamma \backslash G)$. Putting these together, we arrive at the Kuznetrov trace formula for Γ.

THEOREM 2. *Let $\nu \in \mathcal{S}(\mathbb{R}), \mu \in C_c^\infty(\mathbb{R}^\times)$ and set $A(x) = \hat{\nu}(x^2)\mu(\frac{1}{x})$. Then*

$$\sum_{c \in M(\Gamma)} \frac{Kl_\psi(c)}{|c|^2} A(\frac{1}{c}) = \sum_{\pi \subset L^2_{\text{disc}}(\Gamma \backslash G)} c(\pi, \Gamma) \int_{-\infty}^{\infty} A(y) J_\pi(y) d^\times y$$

$$+ \sum_{\text{cusps}} \int_{-\infty}^{\infty} c(\pi(r), \Gamma) \left\{ \int_{-\infty}^{\infty} A(y) J_{\pi(r)}(y) d^\times y \right\} \frac{dr}{4\pi}.$$

Using this formula, we can arrive at an explicit formula for the Kloosterman-Selberg zeta function

$$Z_\Gamma(s) = \sum_{c \in M(\Gamma)} \frac{Kl_\psi(c)}{|c|^{2s}}.$$

and hence its meromorphic continuation. We return to (1.1) and instead of computing $W_{P_f}\left(\begin{pmatrix} 1 & 0 \\ 0 & 1 \end{pmatrix}\right)$ we compute $W_{P_f}\left(\begin{pmatrix} y & 0 \\ 0 & y^{-1} \end{pmatrix}\right)$ and take the Mellin transform. For $Re(s) >> 0$, the formula in Theorem 2 then becomes

$$\sum_{c \in M(\Gamma)} \frac{Kl_\psi(c)}{|c|^{2s}} B\left(\frac{1}{c}\right) = \sum_{\pi \subset L^2_{\text{disc}}(\Gamma \backslash G)} c(\pi, \Gamma) \mathcal{J}(B, \pi, s)$$
$$+ \sum_{\text{cusps}} \int_{-\infty}^{\infty} c(\pi(r), \Gamma) \mathcal{J}(B, \pi(r), s) \frac{dr}{4\pi}$$

where now $\mathcal{J}(B, \pi, s)$ is the Bessel-Mellin transform

$$\mathcal{J}(B, \pi, s) = \int_{-\infty}^{\infty} B(y) J_\pi(y) y^{2s-2} d^\times y$$

and $B(x) = \hat{\nu}(x^2)$ is an even Schwartz function.

If we use Mellin inversion to express B in terms of its Mellin transform $\tilde{B}$, then this formula can be written in a distributional form, namely

$$\tilde{B} \star Z_\Gamma(s) = \tilde{B} \star \sum_{\pi \in L^2_{\text{disc}}(\Gamma \backslash G)} c(\pi, \Gamma) \tilde{J}_\pi(s)$$
$$+ \tilde{B} \star \sum \int c(\pi(r), \Gamma) \tilde{J}_{\pi(r)}(s) \frac{dr}{4\pi}.$$

Now the Mellin transform of the Bessel function $\tilde{J}_\pi(s)$ is the γ-factor $\gamma_\pi(s)$ of the representation π, which is expressible in terms of the L - and ε - functions of Jacquet and Langlands [**3**]. Hence the above gives a distributional equality

$$(3.1) \qquad Z_\Gamma(s) = \sum_{\pi \subset L^2_{\text{disc}}(\Gamma \backslash G)} c(\pi, \Gamma) \gamma_\pi(s) + \sum_{\text{cusps}} \int c(\pi(r), \Gamma) \gamma_{\pi(r)}(s) \frac{dr}{4\pi}.$$

The right hand side of this equation can be shown to converge to an analytic function in a half-plane and in this half-plane the distributional equality becomes an equality of analytic functions. The $\gamma_\pi(s)$ occurring in the right side are expressible in terms of the archimedean L-function $L(\pi, s)$ of the representation (that is, classical Γ-functions) and can be seen to have a meromorphic continuation. This then gives the meromorphic continuation of $Z_\Gamma(s)$ and gives the precise location of its poles in terms of the spectral decomposition of $L^2(\Gamma\backslash G)$.

To make this more precise, let $\pi^0_{\frac{1}{2}+ir}$ denote the unramified principal series representation of $SL_2(\mathbb{R})$ with parameter $\frac{1}{2}+ir$, π^0_r the complimentary series of parameter r. If $\pi \in L^2_{\text{disc}}(\Gamma\backslash G)$, then let $a_1(\pi)$ denote the first Fourier coefficient of the classical modular form of lowest weight corresponding to π, normalized to have square norm one. Let $Z^\sigma_\Gamma(\frac{1}{2}+ir)$ denote the first Fourier coefficient of the Eisenstein series formed at the cusp σ off the representation $\pi^0_{\frac{1}{2}+ir}$. Then formula (3.1) gives the following result.

THEOREM 3. *For $Re(s) > \frac{1}{2}$ we have*

$$\begin{aligned} Z_\Gamma(s) = & \sum_{\substack{\pi \subset L^2_{\text{disc}}(\Gamma\backslash G) \\ \pi \simeq \pi^0_{\frac{1}{2}+ir}}} \frac{|a_1(\pi)|^2}{4(2\pi)^{2s-1} ch(\pi r)} sin(\pi s)\Gamma(s-\frac{1}{2}+ir)\Gamma(s-\frac{1}{2}-ir) \\ & + \sum_{\substack{\pi \subset L^2_{\text{disc}}(\Gamma\backslash G) \\ \pi \simeq \pi^0_r}} \frac{|a_1(\pi)|^2}{4(2\pi)^{2s-1} sin(\pi r)} sin(\pi s)\Gamma(s-1+r)\Gamma(s-r) \\ & - \frac{1}{4\pi}\frac{1}{(2\pi)^{2s-1}}\frac{\Gamma(s)}{\Gamma(1-s)} \\ & - \sum_{\ell=0}^{\infty} \frac{(-1)^\ell(2\ell+1)}{(2\pi)^{2s-1}} \Big(\sum_{c \in M(\Gamma)} \frac{Kl_\psi(c)}{|c|} J_{2\ell+1}\big(\frac{4\pi}{c}\big)\Big) sin(\pi s)\Gamma(s+\ell)\Gamma(s-1-\ell) \\ & + \sum_{\substack{\text{cusps} \\ \sigma}} \frac{sin(\pi s)}{(2\pi)^{2s-1}} \int_{-\infty}^{\infty} |Z^\sigma_\Gamma(\frac{1}{2}+ir)|^2 \Gamma(s-\frac{1}{2}+ir)\Gamma(s-\frac{1}{2}-ir)\frac{dr}{4\pi} \end{aligned}$$

From this, the meromorphic continuation and the location of the poles can be read off rather easily. In his original paper on the sum formula Kuznetsov gave such an expression for $Z_\Gamma(s)$ for $\Gamma = SL_2(\mathbb{Z})$ and this result is a generalization of his formula. Complete results and proofs for this section will appear in a forthcoming paper.

REFERENCES

1. W. Casselman, *Canonical extensions of Harish-Chandra modules to representations of G*, Preprint.
2. J.W. Cogdell, J.-S. Li, I.I. Piatetski-Shapiro and P. Sarnak, *Poincaré series for $SO(n,1)$*, In preparation.

3. R. Godement, *Notes on Jacquet-Langlands Theory*, Lecture Notes, I.A.S..
4. N.V. Kuznetsov, *Petersson's conjecture for cusp forms of weight zero and Linnik's conjecture. Sums of Kloosterman sums*, Math. USSR Sbornik **39** (1981), 299–342.
5. A. Selberg, *On the estimation of Fourier coefficients of modular forms*, Proc. Symp. Pure Math VIII, A.M.S., Providence, R.I. (1965), 1–15.
6. N.R. Wallach, *Asymptotic expansions of generalized matrix entries of representations of real reductive groups. Lie Group Representations I.*, Lecture Notes in Mathematics **1024**, 287–369.

Deformation of Compact Riemann Surfaces Y of Genus p with Distinguished Points $P_1, ..., P_m \in Y$

Dedicated to E. Vesentini to the occasion of his 60th birthday

Gerd Dethloff and Hans Grauert
Mathematisches Institut, Universität Göttingen
Bunsenstr. 3–5
D–3400 Göttingen

1 Introduction

In [G] an idea is given how in very general algebraic spaces X a jetmetric Λ can be constructed which has the following property:
If $C \subset X$ is a local complex curve then $\Lambda|_C$ is a hermitean metric with negative Gaussian curvature K, where K is bounded away from zero independantly of C and X is complete in resp. of this metric. If such a metric exists then X is hyperbolic in the sense of Kobayashi, c.f. [Kob].

The construction in [G] uses families of osculating algebraic curves along C which may be punctured if X is not compact. In this paper we show more generally for families of punctured Riemann surfaces:
The corresponding family of the universal coverings is real bianalytically and on the fibers holomorphically a cartesian product. Especially the family of the hyperbolic metrics on the punctured Riemann surfaces is a real analytic family.

These results are obtained by proving that the so called Fricke coordinates of a (at least real analytic) family of punctured Riemann surfaces (see [Abi]) depend real analytically on the fiber parameter. This result probably was more or less known before by completely different methods, using quasi conformal mappings and the Teichmüller theory (c.f. [Abi], he uses the result that the Fricke coordinates are real analytic coordinates for the Teichmüller space without proof. From this result our results can be obtained quite easily by the theory of Teichmüller spaces). Our proof uses the general theory of deformation of complex spaces in higher dimensions. So it is shown that this meanwhile established theory also applies to the special case of deformation of punctured Riemann surfaces.

We want to thank G. Schumacher for valuable hints for relevant literature.

2 Deformation of Punctured Compact Riemann Surfaces and Infinitesimal Deformation

Assume that X, B are real analytic manifolds and that $\pi : X \to B$ is a surjective, proper, smooth real analytic map.
Further assume that there exists a special atlas $\mathcal{A}$ of charts of X of the following form:

$$(w_1, w_2, z_1, ..., z_{2n}) : U \, (\subset X) \to \mathbb{R}^{2n+2}$$

s.th. for every $x \in U$ $(z_1, ... z_{2n})$ are real analytic coordinates around $\pi(x) \in B$; and s.th. for two such charts the w–components depend holomorphically on each other for fixed z–components.
The atlas $\mathcal{A}$ makes every fiber $X_t := \pi^{-1}(t)$, $t \in B$ a compact complex submanifold of X. We assume that every X_t is a Riemann surface of genus p. Moreover we assume that there are given real analytic cross sections $s_1(t), ..., s_m(t)$ in X over B s.th. for every $t \in B$ the points $s_1(t), ..., s_m(t)$ are disjoint.

Definition 2.1 *X is called a real analytic deformation of Riemann surfaces of genus p with m distinguished points. If additionally X, B are complex manifolds, $\mathcal{A}$ is a holomorphic atlas, and $\pi : X \to B$ and $s_1, ..., s_m$ are holomorphic, then X is called a holomorphic deformation of Riemann surfaces of genus p with m distinguished points.*

In all what follows we mean by real analytic resp. holomorphic deformation these special kinds of deformations.

Next we define the infinitesimal deformations of such deformations in a point $t \in B$. We first assume that the deformation is real analytic. Let $\mathcal{E}_t$ be the tangent bundle defined by the charts of $\mathcal{A}$, restricted to X_t, and $\mathcal{F}_t$ be the subbundle of those tangent vectors along the fibers. Let Θ_t resp. Ψ_t be the sheaves of holomorphic cross sections in $\mathcal{F}_t$ resp. $\mathcal{E}_t$. Further let Θ_t^* be the subsheaf generated by those sections of Θ_t which are zero in $s_1(t), ..., s_m(t)$.
We take a covering $\mathcal{U} = \{U_1, ..., U_l\}$ of open subsets $U_1, ..., U_l$ in X_t s.th. for every $\xi \in T_{B,t}$, there exist fields $\eta_\nu \in H^0(U_\nu, \Psi_t|_{U_\nu})$ with
1) $\pi_*(\eta_\nu) \equiv \xi$
2) η_ν is tangent along the $s_1(t), ..., s_m(t)$.
Then the η_ν, $\nu = 1, ..., l$ define a 0–cochain $\{\eta_\nu\} \in C^0(\mathcal{U}, \Psi_t)$ in the Čech complex defined by the sections of the sheaf Ψ_t and the covering $\mathcal{U}$. It yields a cocycle $d\{\eta_\nu\} \in Z^1(\mathcal{U}, \Theta_t^*)$ which, by passing to the limit, yields a cohomology class $\rho_t(\xi) \in H^1(X_t, \Theta_t^*)$. This is uniquely determined and $\rho_t : T_{B,t} \to H^1(X_t, \Theta_t^*)$ is a real linear map.
If the deformation is holomorphic then we define completely analogous using the holomorphic instead of the mixed real analytic and holomorphic structure. We get a complex linear map $\rho_t : T_{B,t} \to H^1(X_t, \Theta_t^*)$.

Definition 2.2 *The map $\rho_t : T_{B,t} \to H^1(X_t, \Theta_t^*)$ is called the infinitesimal deformation of the real analytic resp. holomorphic deformation.*

Definition 2.3 *A real analytic resp. holomorphic deformation is called stable in* $t \in B$ *(with resp. to* $s_1, ..., s_m$ *) if it is a cartesian product over the double point in* t *, i.e. if there are finitely many charts* $(w^{(i)}, z) : U_i \to \mathbb{R}^{2n+2}$, $i = 1, ..., l$ *out of the atlas* $\mathcal{A}$ *with:*
1) $X_t \subset \bigcup_{i=1}^{l} U_i$
2) $\frac{\partial s_j}{\partial z_k}(t) = 0$ *for* $j = 1, ..., m,$ z_k *being any component of* z
3) For any coordinate transformation $w^{(j)} = w^{(j)}(w^{(i)}, z)$ *we have* $\frac{\partial w^{(j)}}{\partial z_k}(t) = 0.$

We have the following

Theorem 2.4 *A real analytic resp. holomorphic deformation is stable in* $t \in B$ *with resp. to* $s_1, ..., s_m$ *iff* $\rho_t(\xi) \equiv 0$ *in* $T_{B,t}$.

To prove that, it is first shown by using the long exact cohomology sequence belonging to the short exact sheaf sequence

$$0 \to \Theta_t^* \to \Psi_t \to \Psi_t/\Theta_t^* \to 0$$

that $\rho_t(\xi) \equiv 0$ iff for all $\xi \in T_{B,t}$ there exists a field $\eta \in \Psi_t(X_t)$ with $\pi_*(\eta) \equiv \xi$. Now it can easily be shown that the existence of such fields is equivalent with the existence of special charts in $\mathcal{A}$ like in Definition 2.3. For more details c.f. [KS,I]. □

3 Existence of Semiuniversal Deformation

Definition 3.1 *A real analytic (resp. holomorphic) deformation* $\pi : X \to B$ *is called semiuniversal in* $t \in B$ *if:*
1) If $\psi : Z \to G$ *is an other real analytic (resp. holomorphic) deformation with* $0 \in G$ *and* $Z_0 = X_t$ *then there is a real analytic (resp. holomorphic) map* $\alpha : U(0) \to B$ *with* $\alpha(0) = t$ *s.th.* $X \circ \alpha = X \times_\alpha U$ *is isomorphic to* $Z|_U$ *under an isomorphism which is on* $Z_0 = X_t$ *the identity.*
2) The total derivative $d\alpha : T_{G,0} \to T_{B,t}$ *is uniquely determined.*

The semiuniversal deformation is uniquely determined always.
Remark: It is possible to prove that it is universal, i.e. that α itself is uniquely determined.

Lemma 3.2 *If* Y *is a compact Riemann surface of genus* p *with distinguished points* $P_1, ..., P_m \in Y$ *there exists a holomorphic semiuniversal deformation* $\pi : X \to B \subset \mathbb{C}^n$ *with* $Y = X_0$, $P_i = s_i(0), i = 1, ..., m$ *and* $n = \dim_{\mathbb{C}} H^1(X_0, \Theta_0^*)$.

Proof: We have $H^2(Y, \Theta_t) = 0$. So if we forget about the distinguished points in Y for a moment there exists a seminuiversal deformation of compact Riemann surfaces (in the ordinary sense) $\pi' : X' \to B'$ with $X'_0 = Y$ and bijective infinitesimal deformation $\rho_0 : T_{B',0} \to H^1(X'_0, \Theta_0)$, cf. [KNS], [KS,II].

If we now additionally have holomorphic cross sections $s_1, ..., s_m$ through $P_1, ..., P_m$, we get $r := \max\{0, m - \max\{3-2p, 0\}\}$ additional deformation parameters, which are independant from the others. So we have a holomorphic deformation (again of our kind, cf. Definition 2.1) $\pi : X \to B' \times \mathbb{C}^r$ s.th. $X_0 = Y$ and $\rho_0 : T_{B' \times \mathbb{C}^r, 0} \to H^1(X_0, \Theta_0^*)$ again is bijective. The last two statements can be proved by applying the theorem of Riemann–Roch. Now we still have to show that this family is complete in $t = 0$ since then, by the bijectivity of ρ_0, it is semiuniversal in the zero point . This can be proved like in [KS,II], but we also can get this immediately by using a general theorem stated in [Fl]. □

Lemma 3.3 *If* $\pi : X \to B$ *is semiuniversal in* $t \in B$ *for holomorphic deformations, then it there also is semiuniversal for real analytic deformations.*

Proof: Assume that $\Psi : Z \to G$ is a real analytic deformation with $0 \in G$ and $Z_0 = X_t$. We may assume that $G \subset \mathbb{R}^l$ with coordinates $x_1, ..., x_l$. We take a small domain $\tilde{G} \subset \mathbb{C}^l$ with complex coordinates $\tilde{x_1}, ..., \tilde{x_l}$ s.th. $G = \tilde{G} \cap \mathbb{R}^l$ and everything can be extended holomorphically onto $\tilde{G}$: Thus we get a holomorphic deformation $\tilde{\Psi} : \tilde{Z} \to \tilde{G}$ and disjoint cross sections $\tilde{s_1}, ..., \tilde{s_m}$. Since $\pi : X \to B$ is semiuniversal for holomorphic families we obtain a holomorphic map $\tilde{\alpha} : U(0) \to B$ with $\tilde{\alpha}(0) = t$ s.th. $\tilde{Z}|_U = X \circ \tilde{\alpha}$ and $d\tilde{\alpha}$ is uniquely determined.
By restricting $\alpha = \tilde{\alpha}|_{U \cap G}$ we have $Z|_{U \cap G} = X \circ \alpha$. Since $d\tilde{\alpha}$ maps infinitesimal deformation onto the same infinitesimal deformation and this remains true under restriction of $T_{\tilde{G},0}$ to $T_{G,0}$ and $\tilde{\alpha}$ to α the map $d\alpha$ is uniquely determined. □

4 Real Analytic Triviality of the Family of the Universal Coverings

We now wish to construct a real analytic deformation $\Psi : Z \to G$ with $Z_0 = Y$, where Y is again a given compact Riemann surface with distinguished points $P_1, ..., P_m$, and with bijective $\rho_0 : T_{G,0} \to H^1(Z_0, \Theta_0^*)$ and on G real analytic Fricke coordinates (c.f. [Abi] or see below). We need some preparations:
We put $Y' = Y \setminus \{P_1, ..., P_m\}$ and assume from now on $p \geq 2$ or $p = 1$ and $m \geq 1$ or $p = 0$ and $m \geq 3$. Then the universal covering of Y' is the upper half plane H, and Y' can be represented as the quotient of H by a Fuchsian group Γ which is finitely generated and which acts fixpointfree and properly discontinuous on H. Let Π_1 be the fundamental group of Y'. It is generated by loops $\alpha_1, \alpha_1^*, ..., \alpha_p, \alpha_p^*, \beta_1, ..., \beta_m$ with the following properties:
1) The intersection number of α_ν, α_ν^* is 1.
2) All the other loops only have the base point in common. Their intersection number is zero.
If $m = 0$ then the loops $\alpha_1, \alpha_1^*, ..., \alpha_p, \alpha_p^*$ generate Π_1. There is exactly one relation

$$\alpha_1 \cdot \alpha_1^* \cdot \alpha_1^{-1} \cdot (\alpha_1^*)^{-1} \cdot ... \cdot \alpha_p \cdot \alpha_p^* \cdot \alpha_p^{-1} \cdot (\alpha_p^*)^{-1} = 1.$$

If $m \neq 0$ Π_1 is generated by $\alpha_1, ..., \beta_m$ and is free with the $2p+m-1$ generators $\alpha_1, ..., \beta_{m-1}$. We have an isomorphism $\chi : \Pi_1 \to \Gamma$. Let

$$\chi(\alpha_\nu) = \frac{a_\nu z + b_\nu}{c_\nu z + d_\nu} \quad \chi(\alpha_\nu^*) = \frac{a_\nu^* z + b_\nu^*}{c_\nu^* z + d_\nu^*} \quad \chi(\beta_\nu) = \frac{\tilde{a_\nu} z + \tilde{b_\nu}}{\tilde{c_\nu} z + \tilde{d_\nu}}$$

where the determinants always are one and all coefficients are real. Since the $\chi(\alpha_\nu), \chi(\alpha_\nu^*)$ are hyperbolic, the $\chi(\beta_\nu)$ are parabolic ([Abi,p.42]), the relation has to be satisfied and H has a three parameter group we have $2(3p-3+m)$ real parameters from which all the coefficients can be recovered ([Abi,p.64]). (Both references don't depend on the Teichmüller theory or quasi conformal mappings.)
Now we can construct the desired real analytic deformation:
We take $G \subset \mathbb{C}^{3p-3+m}$ with $0 \in G$ and define $2p+m$ automorphisms

$$\frac{a_\nu(t)z + b_\nu(t)}{c_\nu(t)z + d_\nu(t)} \quad \frac{a_\nu^*(t)z + b_\nu^*(t)}{c_\nu^*(t)z + d_\nu^*(t)} \quad \frac{\tilde{a_\nu}(t)z + \tilde{b_\nu}(t)}{\tilde{c_\nu}(t)z + \tilde{d_\nu}(t)} \tag{1}$$

of H with coefficients depending real analytically on t s.th. the coefficients satify our conditions (hence we only may define $2(3p-3+m)$ coefficient functions, the others are fixed by these conditions), the quotient of H by these automorphisms for $t=0$ yields Y', and s.th. for no $\xi \in T_{G,0}$, $\xi \neq 0$ the derivatives of the coefficient functions in ξ–direction vanish simultaneously. Let Γ_t be the group generated by the automorphisms of (1) for $t \in G$ and $\Gamma_G := \Gamma_t, t \in G$. Then (possibly after having made G smaller) for all $t \in G$ the group Γ_t is properly discontinuous, fixpointfree and finitely generated:
Let D_0 be a fixed fundamental region of Γ_0. If G is small enough one can (possibly after having changed the generators of Γ_G for a moment), by the action of 'one letter words' $w(t) \in \Gamma_G$ in a small neighborhood of D_0, find 'fundamental regions' D_t which depend real analytically on t. Then we always have $w(t)(\dot{D}_t) \cap \dot{D}_t = \emptyset$ (where $\dot{D}_t$ denotes the open kernel of D_t), for longer words $g(t) \in \Gamma_G$, of course, that equation needn't hold. H is covered by the translates of D_t given by the elements of Γ_t for every $t \in G$. We have to show that the open kernels of those translates are disjoint.
Assume that for some t_0 two of them aren't disjoint. Then there also exists a $g_0(t_0) \in \Gamma_{t_0}$ with $\dot{D}_{t_0} \cap g_0(t_0)(\dot{D}_{t_0}) \neq \emptyset$. Let $\Phi : [0,1] \to G$ with $\Phi(0) = t_0$, $\Phi(1) = 0$ be any curve. If $g_0(0)(\dot{D}_0) \cap \dot{D}_0 = \emptyset$ we obtain by multiplying $m \geq 1$ 'letters' w_j from the left:
There exist open intervals $I_0, ..., I_l \subset \mathbb{R}$ with $\bigcup_{i=0}^{l} I_i \supset [0,1]$, $0 \in I_0$, $1 \in I_l$, $[0,1] \cap I_i \neq \emptyset$ for all $i = 0, ..., l$ and $g_i(t) := w_i(t) \cdot ... \cdot w_1(t) \cdot g_0(t) \in \Gamma_G$, $i = 1, ..., l$ with $g_i(t)(\dot{D}_t) \cap \dot{D}_t \neq \emptyset$ for $t \in \Phi(I_i \cap [0,1])$. We have $g_l(0) = id$ since Γ_0 was properly discontinuous and fixpointfree. Since the relations are kept while deforming Γ_0 to Γ_t we have $g_l = id$ on G. But then we have $g_{l-1}(t)(\dot{D}_t) \cap \dot{D}_t = w_l^{-1}(t)(\dot{D}_t) \cap \dot{D}_t = \emptyset$ by construction of D_t for all $t \in G$, which is a contradiction. So we have $g_0(0)(\dot{D}_0) \cap \dot{D}_0 \neq \emptyset$, hence $g_0(0) = id$ and then by the same argument as above we have $g_0 = id$ on G which proves our assertion.
Therefore $(H \times G)/\Gamma_G$ is a well defined family of Riemann surfaces with $H/\Gamma_0 = Y'$ on which $H \times G$ canonically yields real analytic and on the fibers holomorphic charts.

In order to get a real analytic deformation $\Psi : Z \to G$ like in Definition 2.1 from $(H \times G)/\Gamma_G$, we have to 'fill in the punctures'. If S resp. S' denote the underlying topological spaces of Y resp. of $Y' = Y \backslash \{P_1, ..., P_m\}$ and $s_i(t) := (P_i, t) \in S \times G$, $i = 1, ..., m$, that means that we can find real analytic and on the fibers holomorphic charts (with resp. to those charts on $(H \times G)/\Gamma_G$, the underlying topological space of which is by construction $S' \times G$ since the real analytic and on the fibers complex structure of $(H \times G)/\Gamma_G$ can be given through the automorphism $\chi_t : \Pi_1(S') \to \Gamma_t$) around all points $s_i(t) \in S \times G$, $t \in G, i = 1, ..., m$, with the following properties:
All those charts give an atlas $\mathcal{A}$ (cf. Definition 2.1) on $S \times G$ in such a way that $(H \times G)/\Gamma_G = Z \setminus \{s_1(G), ..., s_m(G)\}$, where Z denotes $S \times G$ with the structure given by $\mathcal{A}$, the fibers of Z are compact Riemann surfaces of genus p, and the $s_i(t)$, $t \in G$ are real analytic cross sections in Z. Let $Q = s_i(t_0)$ be arbitrary. Let V be a small neighborhood of Q in $S \times G$. Then by the isomorphism $\chi_t : \Pi_1(S') \to \Gamma_t$ the deck transformation $\beta_i(t)$ covers small simple loops in $S' \times \{t\}$ around the 'puncture' $s_i(t)$. $\beta_i(t)$ has exactly one fixed point $F(t) \in \partial\Delta$ (where we identify H and Δ through a fixed biholomorphic map) which depends real analytically on $t \in G$. So there exists a family of rotations in the fibers depending real analytically on t which maps $H \times G$ real bianalytically and on the fibers holomorphically onto itself s.th. the border points $F(t)$ are mapped to infinity. Now $\beta_i(t)$ is a translation $z \to z + b(t)$, where $b(t)$ depends real analytically on t. So the automorphism $(w, z) \to (b(t)^{-1} w, z)$ on $H \times G$ again is real bianalytic and on the fibers holomorphic. At last we map $H \times G$ to $\Delta \times G$ by $(w, z) \to (e^{iw}, z)$. Then the invers image of $V(Q)$ in $\Delta \times G$ is the intersection of an open neighborhood $W((0, t_0)) \subset \Delta \times G$ with $(\Delta \setminus \{0\}) \times G$ which, by construction and the properties of $\beta_i(t)$, is mapped real bianalytically and on the fibers holomorphically to $V \setminus s_i(G)$.
So we have constructed the desired charts and hence the real analytic deformation, the Fricke coordinates of which, i.e. the $2(3p - 3 + m)$ coefficient functions in (1) which weren't given by the restructions, depend by construction real analytically on G.
Since by the theorem of Riemann–Roch we have $\dim_{\mathbb{C}} H^1(Z_0, \Theta_0^*) = 3p - 3 + m$, we only have to show that the map $\rho_0 : T_{G,0} \to H^1(Z_0, \Theta_0^*)$ is injective.
Let $\xi = \sum_{\nu=1}^{2(3p-3+m)} a_\nu \frac{\partial}{\partial t_\nu} \in T_{G,0}$, $\xi \neq 0$, $L = \{t = (a_1, ..., a_{2(3p-3+m)})s,\ s \in \mathbb{R}\} \cap G$ be the real line in ξ–direction, and 0_1 be the double point in $0 \in L$. If we restrict the transformations (1) to 0_1 their deformation is by construction not zero. Hence the restriction of $Z|_L$ to 0_1 has nonvanishing infinitesimal deformation: The complex structure on the fibers is parametrized by the elements of Γ_G up to the automorphism group of H. Since in the family (1) of generators of Γ_G the influence of this automorphism group already has been thrown out by a reduction of the number of free parameters (by our conditions for the coefficient functions) this means that the complex structures of the fibers are 'changing over 0_1'. Hence $Z|_L$ cannot be stable over 0_1, since otherwise there would exist real analytic and on the fibers biholomorphic coordinates which would make $Z|_L$ a cartesian product over 0_1, but then the complex structures of the fibers couldn't 'change over 0_1'. Hence the infinitesimal deformation of $Z|_L$ in the zero point isn't zero.
Since $\xi \in T_{G,0}, \xi \neq 0$ was arbitrary this means that the infinitesimal deformation of

$\Psi : Z \to G$ is injective in the zero point. So our deformation $\Psi : Z \to G$ has all desired properties.

We now can prove the following:

Theorem 4.1 *Let $\pi' : X' \to B'$ be a real analytic deformation (like in Definition 2.1). Then its Fricke coordinates (more exactly those of $\pi' : X' \setminus \{s'_1, ..., s'_m\} \to B'$) depend real analytically on B'. Especially the universal covering $\hat{X}'$ of $X' \setminus \{s'_1, ..., s'_m\}$ is real bianalytically and on the fibers holomorphically equivalent to $H \times B'$, where H denotes the upper half plane, and the hyperbolic metric on the fibers of $X' \setminus \{s'_1, ..., s'_m\}$ depends real analytically on the fiber parameter $t \in B'$.*

Proof: Let $t \in B'$ be arbitrary. Then with Lemma 3.2 there exists a holomorphic semiuniversal deformation $\pi : X \to B \subset \mathbb{C}^n$ with $X_0 = X'_t$ and $n = \dim_{\mathbb{C}} H^1(X_0, \Theta_0^*)$. From Lemma 3.3 we know that $\pi : X \to B$ is also semiuniversal for real analytic deformations. Together with our preceding construction this means:

1) There exists a real analytic mapping $\alpha' : U(t)(\subset B') \to B$ s.th. $X'|_U \cong X \circ \alpha'$.
2) There exists a real analytic deformation $\Psi : Z \to G$ like constructed above with $Z_0 = X_0$ and with an in the zero point bijective infinitesimal deformation. There further exists a a real analytic map $\alpha : V(0) \to B$ with $Z|_V \cong X \circ \alpha$.

Since $d\alpha$ maps infinitesimal deformation to the same infinitesimal deformation and the infinitesimal deformation of $\Psi : Z \to G$ in 0 is bijective, $d\alpha$ is injective. Since $\dim_{\mathbb{C}} B = \dim_{\mathbb{C}} H^1(X_0, \Theta_0^*) = \dim_{\mathbb{C}} G$ the map $d\alpha$ is bijective. By the invers function theorem $\alpha^{-1} : B \to G$ exists locally around $0 \in B$. So we have $X \cong Z$ locally around 0 through the real bianalytic map α. Especially the Fricke coordinates of X depend real analytically on B since the Fricke coordinates of Z depend real analytically on G, both around the zero point. Since by 1) the family $\pi' : X' \to B'$ can be obtained by lifting from $\pi : X \to B$ the Fricke coordinates of X' depend real analytically on B'.

The other both assertions are an immediate consequence of this fact. □

Remark: In the cases of deformations with $p = 0$, $m \leq 2$ and $p = 1$, $m = 0$ (c.f. Definition 2.1) we cannot deform by using a Fuchsian group. In the case $p = 0$, $m \leq 2$ the deformation is trivial, and for $p = 1$, $m = 0$ the deformation is parametrized by the upper half plane H.

References

– [Abi]: Abikoff, W.: The Real Analytic Theory of Teichmüller Spaces. Berlin, Heidelberg, New York. Springer 1980

– [Fl]: Flenner, H.: Über Deformationen holomorpher Abbildungen. Habilitationsschrift. Osnabrück 1978

– [G]: Grauert, H.: Jetmetriken und hyperbolische Geometrie. Math. Z. **200** (1989)

– [KNS]: Kodaira, K., Nierenberg, L., Spencer, D.C.: On the Existence of Deformations of Complex Analytic Structures. Ann. of Math. **68**, 450–459 (1958)

– [KS,I]: Kodaira, K., Spencer, D.C.: On Deformation of Complex Analytic Structure I, II. Ann. of Math **67**, 328–468 (1958)

– [KS,II]: Kodaira, K., Spencer, D.C.: A Theorem of Completeness for Complex Analytic Fiber Spaces. Acta Math. **100**, 281–294 (1958)

– [Kob]: Kobayashi, S.: Intrinsic Distances, Measures and Geometric Function Theory. Bull. Am. Math. Soc. **82**, 357–416 (1976)

On Moduli of Vector Bundles

Shoshichi Kobayashi
Department of Mathematics, University of California, Berkeley

1. Introduction.

We wish to discuss here moduli spaces of simple vector bundles on a compact Kähler manifold M from differential geometric viewpoints, placing emphasis on the case where M is symplectic Kähler.

In order to explain the construction of such moduli spaces, we consider first an analogous construction of the moduli space of complex structures on a differentiable manifold M. Let $\mathcal{A}(M)$ be the set of almost complex structures J on M. Each almost complex structure J gives rise to a decomposition

$$d = d' + d''$$

of exterior differentiation d. Then J is integrable if and only if $d'' \circ d'' = 0$. Let $\mathcal{C}(M)$ be the set of integrable almost complex structures on M. The group $\mathcal{D}(M)$ of diffeomorphisms of M acts on $\mathcal{A}(M)$ and $\mathcal{C}(M)$. Then the moduli space of complex structures on M is given by $\mathcal{C}(M)/\mathcal{D}(M)$.

Now, fix a C^∞ complex vector bundle E of rank r over a compact complex manifold M. A compatible almost complex structure in E is given by a differential operator D'' sending each section of E to an E-valued (0,1)-form on M and satisfying conditions similar to those for covariant differentiation. The integrability condition is given by $D'' \circ D'' = 0$. We denote the set of integrable D'' by $\mathcal{H}''(E)$. As an analogue to the group of diffeomorphisms of M we consider this time the group $GL(E)$ of automorphisms of E (inducing the identity transformation on the base manifold M). Then the group $GL(E)$ acts on $\mathcal{H}''(E)$, and the space of orbits, $\mathcal{H}''(E)/GL(E)$, is the moduli space of holomorphic structures in E.

However, in order to provide this moduli space with a good complex analytic structure, we have to restrict our consideration to simple holomorphic structures. A holomorphic vector bundle is said to be simple if it admits no (sheaf) endomorphisms other than scalar multiplications. The set of integrable D'' that give rise to simple holomorphic structures is denoted $\tilde{\mathcal{H}}''(E)$. The moduli space $\tilde{\mathcal{M}}(E) = \tilde{\mathcal{H}}''(E)/GL(E)$ of simple holomorphic structures in E is a (possibly non-Hausdorff) complex analytic space (in generally, not reduced).

By considering only stable holomorphic structures in E or Einstein-Hermitian structures in E we obtain a Hausdorff moduli space. Given an Hermitian structure h in a C^∞ complex vector bundle E over a compact Kähler manifold M, the moduli space $\tilde{\mathcal{M}}(E, h)$ of irreducible Einstein-Hermitian connections in (E, h) carries a natural Kähler metric while $\tilde{\mathcal{M}}(E)$ may not. In spite of apparent advantages of $\tilde{\mathcal{M}}(E, h)$ over $\tilde{\mathcal{M}}(E)$, we consider here mainly the latter since it is much easier to keep track of various holomorphic objects on $\tilde{\mathcal{M}}(E)$ than those on $\tilde{\mathcal{M}}(E,h)$. It is also much easier to explain the algebraic concept of simple vector bundle than that of Einstein-Hermitian vector bundle. This is somewhat analogous to the situation for a compact homogeneous Kähler manifold V which can be expressed either as a homogeneous space $V = K/L$ of a compact Lie group K of isometries or as a homogeneous space $V = G/H$ of a complex Lie group G of holomorphic transformations; while metric properties of V can be seen more easily from K/L, it is often more convenient to use G/H in order to study holomorphic objects on V or algebraic properties of V. While K/L and G/H are diffeomorphic in a natural way, $\tilde{\mathcal{M}}(E,h)$ is open in $\tilde{\mathcal{M}}(E)$ and does not exactly agree with $\tilde{\mathcal{M}}(E)$. However, it is often sufficient to study $\tilde{\mathcal{M}}(E)$ in order to understand $\tilde{\mathcal{M}}(E,h)$.

We generally use the notation of [9].

2. Holomorphic Structures of Vector Bundles. [5,6,9,10]

We fix an n-dimensional compact Kähler manifold M and a differentiable complex vector bundle E of rank r over M. We wish to consider the set of holomorphic vector bundle structures in E compatible with the given complex structure of M and the complex vector bundle structure of E. Let $A^{p,q}(E)$ be the space of C^∞ differential (p,q)-forms on M with values in E, and set

$$A^r(E) = \sum_{p+q=r} A^{p,q}(E).$$

If E is a trivial line bundle, these spaces are denoted simply $A^{p,q}$ and A^r.

Let $\mathcal{D}''(E)$ denote the set of $\mathbf{C}$-linear maps

$$D'' : A^0(E) \to A^{0,1}(E)$$

satisfying

$$D''(fs) = (d''f)s + f \cdot D''s \qquad \text{for} \quad s \in A^0(E), f \in A^0.$$

Every D'' extends uniquely to a $\mathbf{C}$-linear map

$$D'' : A^{p,q}(E) \to A^{p,q+1}(E)$$

satisfying

$$D''(\psi\sigma) = d''\psi \wedge \sigma + (-1)^{r+s}\psi \wedge D''\sigma \qquad \text{for} \quad \sigma \in A^{p,q}(E), \psi \in A^{r,s}.$$

The total space of a C^∞ complex vector bundle E over a complex manifold M has no *natural* almost complex structure. But each $D'' \in \mathcal{D}''(E)$ gives rise, in a natural manner, to an almost complex structure on E that is compatible with the complex structure of M and the complex vector space structure of the fibers. Then the almost complex structure is integrable if and only if D'' satisfies the integrability condition:

$$D'' \circ D'' = 0.$$

The holomorphic vector bundle defined by E and an integrable D'' will be denoted $E^{D''}$. Then an integrable D'' is nothing but d'' of the holomorphic vector bundle $E^{D''}$. Let $\mathcal{H}''(E) \subset \mathcal{D}''(E)$ be the subset consisting of integrable D''.

We know that $\mathcal{D}''(E)$ is a complex affine space (of infinite dimension). In fact, if we fix $D_0'' \in \mathcal{D}''(E)$, then for every $D'' \in \mathcal{D}''(E)$ the difference $\alpha = D'' - D_0''$ is a map

$$\alpha : A^0(E) \to A^{0,1}(E)$$

which is linear over A^0, and α can be regarded as an element of $A^{0,1}(End(E))$. Conversely, for any element $\alpha \in A^{0,1}(End(E)), D'' = D_0'' + \alpha$ is an element of $\mathcal{D}''(E)$. Thus, once D_0'' is chosen and fixed as the origin, $\mathcal{D}''(E)$ can be identified with the complex vector space $A^{0,1}(End(E))$.

However, $\mathcal{H}''(E)$ is not an affine subspace of $\mathcal{D}''(E)$. In fact, the formula

$$D'' \circ D'' = D_0'' \circ D_0'' + D_0'' \circ \alpha + \alpha \circ D_0'' + \alpha \wedge \alpha$$

shows that when D_0'' is integrable, the integrability condition for D'' is given by

$$D_0''\alpha + \alpha \wedge \alpha = 0,$$

which is quadratic in α. Let $GL(E)$ denote the group of C^∞ bundle automorphisms of E (which induce the identity transformation on M). The space of C^∞ sections of the endomorphism bundle $End(E) = E \otimes E^*$ will be denoted $\mathbf{gl}(E)$ and is considered as the Lie algebra of $GL(E)$.

The group $GL(E)$ acts on $\mathcal{D}''(E)$ by

$$D'' \mapsto D''^f = f^{-1} \circ D'' \circ f = D'' + f^{-1}d''f,$$

where $D'' \in \mathcal{D}''(E)$ and $f \in GL(E)$. Then $GL(E)$ sends $\mathcal{H}''(E)$ into itself. Two holomorphic structures $D_1'', D_2'' \in \mathcal{H}''(E)$ of E are considered *equivalent* if they are in the same $GL(E)$-orbit. The space $\mathcal{M}(E) = \mathcal{H}''(E)/GL(E)$ of $GL(E)$-orbits with the C^∞-topology is the moduli space of holomorphic structures in E.

Let $[D'']$ denote the point of $\mathcal{M}(E)$ represented by $D'' \in \mathcal{H}''(E)$. In order to describe the tangent space to $\mathcal{M}(E)$ at $[D'']$, let

$$D_t'' = D'' + \alpha_t''$$

be a curve in $\mathcal{H}''(E)$, where $\alpha_t'' \in A^{0.1}(End(E))$ and $\alpha_0'' = 0$. Then

$$D''\alpha_t'' + \alpha_t'' \wedge \alpha_t'' = 0.$$

Differentiating the above equation with respect to t at $t = 0$, we obtain

$$D''\alpha'' = 0,$$

where $\alpha'' = \partial_t \alpha_t''|_{t=0}$.

If D_t'' is obtained by a 1-parameter family of transformations $f_t \in GL(E)$, i.e.,

$$D_t'' = f_t^{-1} \circ D'' \circ f_t$$

with $f_0 = I_E$, then

$$D_t'' = D'' + \alpha_t'', \quad \text{where} \quad \alpha_t'' = f_t^{-1}D''f_t.$$

If we set

$$\alpha'' = \partial_t \alpha''_t|_{t=0}, \qquad f = \partial_t f_t|_{t=0},$$

then

$$\alpha'' = D'' f.$$

Hence, the tangent space $T_{[D'']}(\mathcal{M}(E))$ is given by

$$H^{0,1}(M, End(E^{D''})) = \frac{\{\alpha'' \in A^{0,1}(End(E)); D''\alpha'' = 0\}}{\{D''f; f \in A^0(End(E))\}}$$

provided that $\mathcal{M}(E)$ is nonsingular at $[D'']$. (Here, $H^{0,1}(M, End(E^{D''}))$ is the (0,1)-th d''-cohomology of the bundle $End(E^{D''})$ in the sense of Dolbeault).

Every holomorphic vector bundle $E^{D''}$ admits nonzero endomorphisms, namely $cI_E, c \in$ **C**. If these are the only endomorphisms of $E^{D''}$, i.e., if $H^0(M, \mathcal{O}(End(E^{D''}))) = \mathbf{C}$, then $E^{D''}$ is said to be simple. It is convenient to introduce the subbundle $End^0(E^{D''})$ of $End(E^{D''})$ consisting of trace-free endomorphisms. Then $E^{D''}$ is simple if and only if

$$H^0(M, \mathcal{O}(End^0(E^{D''}))) = 0.$$

Let $\tilde{\mathcal{H}}''(E)$ be the subset of $\mathcal{H}''(E)$ consisting of D'' such that $E^{D''}$ is simple, and let

$$\tilde{\mathcal{M}}(E) = \tilde{\mathcal{H}}''(E)/GL(E).$$

Then we have

Theorem (2.1). *The moduli space $\tilde{\mathcal{M}}(E)$ of simple holomorphic structures on E is a (possibly non-Hausdorff and non-reduced) complex analytic space. It is nonsingular at $[D'']$ if $H^2(M, \mathcal{O}(End^0(E^{D''}))) = 0$, and its tangent space at $[D'']$ is isomorphic to $H^1(M, \mathcal{O}(End(E^{D''})))$.*

Even if $[D'']$ is a singular point, $H^1(M, \mathcal{O}(End(E^{D''})))$ is still isomorphic to the Zariski tangent space of $\tilde{\mathcal{M}}(E)$ at $[D'']$.

Remark. Fix an Hermitian structure h in E. Let $\mathcal{D}(E,h)$ denote the set of connections $D = D' + D''$ in E preserving $h, \mathcal{H}(E,h)$ the subset of consisting of D such that $D'' \circ D'' = 0, \mathcal{H}^{\mathcal{E}}(E,h)$ the set of Einstein-Hermitian connections, and $\tilde{\mathcal{H}}^{\mathcal{E}}(E,h)$ the set of irreducible Einstein-Hermitian connections. Let $U(E,h)$ denote the subgroup of $GL(E)$ consisting of automorphisms of E preserving h. We set $\mathcal{M}(E,h) = \mathcal{H}^{\mathcal{E}}(E,h)/U(E,h)$ and $\tilde{\mathcal{M}}(E,h) = \tilde{\mathcal{H}}^{\mathcal{E}}(E,h)/U(E,h)$. Then the moduli space $\tilde{\mathcal{M}}(E,h)$ of irreducible Einstein-Hermitian connections is Hausdorff and open in $\tilde{\mathcal{M}}(E)$ and carries a natural Kähler metric (on its nonsingular part). For details on Einstein- Hermitian connections, see [9].

If L is a trivial line bundle over M, then the moduli space $\mathcal{M}(L)$ is by definition the Picard variety $Pic^0(M)$ of M.

The group $Pic^0(M)$ acts on $\tilde{\mathcal{M}}(E)$ essentially freely. Fixing a holomorphic structure D'' in a C^∞ complex vector bundle E, we set

$$\mathcal{E} = E^{D''}.$$

Then each element $\mathcal{L} \in Pic^0(M)$ gives rise to a holomorphic vector bundle $\mathcal{E} \cdot \mathcal{L}$ with the same underlying C^∞ complex vector bundle E. Clearly, $\mathcal{E}$ and $\mathcal{E} \cdot \mathcal{L}$ have the same associated projective bundle, i.e., $\mathbf{P}(\mathcal{E}) = \mathbf{P}(\mathcal{E} \cdot \mathcal{L})$.

Conversely, if two holomorphic vector bundles $\mathcal{E}$ and $\mathcal{E}'$ have an isomorphic assocaited projective bundle $\mathbf{P}(\mathcal{E}) \cong \mathbf{P}(\mathcal{E}')$, then $\mathcal{E}' \cong \mathcal{E} \cdot \mathcal{L}$ for some holomorphic line bundle $\mathcal{L}$. If $\mathcal{E}$ and $\mathcal{E}'$ have the same underlying C^∞ vector bundle E, then this line bundle $\mathcal{L}$ is in the Picard variety $Pic^0(M)$.

It may happen that $\mathcal{E} \cdot \mathcal{L}$ is isomorphic to $\mathcal{E}$. Following Mukai, we set

$$\Sigma^0(\mathcal{E}) = \{\mathcal{L} \in Pic^0(M); \mathcal{E} \cdot \mathcal{L} \cong \mathcal{E}\}.$$

If $\mathcal{L} \in \Sigma^0(\mathcal{E})$, then $det(\mathcal{E}) \cong det(\mathcal{E}) \cdot \mathcal{L}^r$. Hence, $\Sigma^0(\mathcal{E})$ is contained in the r-torsion of $Pic^0(M)$. It follows that the moduli space $\mathcal{M}(E)$ is fibered by $Pic^0(M)/\Sigma^0(\mathcal{E})$.

On the other hand, by fixing a simple holomorphic structure $D''_o \in \tilde{\mathcal{H}}''(E)$ and setting $\mathcal{E}_o = E^{D''_o}$, we can define a fibering $\mathcal{M}(E) \to Pic^0(M)$ by

$$\mathcal{E} \mapsto det(\mathcal{E}) \cdot det(\mathcal{E}_o)^{-1}.$$

We defined two complementary fiberings of $\tilde{\mathcal{M}}(E)$. The decomposition

$$H^1(M, \mathcal{O}(End(\mathcal{E}))) \cong H^1(M, \mathcal{O}(End^0(\mathcal{E}))) + H^1(M, \mathcal{O})$$

defines a splitting of the tangent bundle of $\tilde{\mathcal{M}}(E)$ into two complementary subbundles. These subbundles correspond to the two fiberings above.

Remark. On $\tilde{\mathcal{M}}(E, h)$, the decomposition above defines two parallel distributions of $\tilde{\mathcal{M}}(E, h)$.

The following construction is consistent with the second fibering above.

Each nonzero holomorphic 1-form φ on M induces, in a natural way, a nonzero closed holomorphic 1-form $\tilde{\varphi}$ on $\tilde{\mathcal{M}}(E)$.

For each tangent vector $\alpha \in A^{0,1}(End(E))$ at $D'' \in \mathcal{D}''(E)$, we set

$$\hat{\varphi}(\alpha) = \int_M \varphi \wedge tr(\alpha) \wedge \Phi^{n-1},$$

where Φ is the Kähler form of M. Then $\hat{\varphi}$ is a holomorphic 1-form on $\mathcal{D}''(E)$. Its restriction to $\mathcal{H}''(E)$ projects down to a holomorphic 1-form $\tilde{\varphi}$ on $\tilde{\mathcal{M}}(E)$.

In fact, we show (i) that if α comes from the infinitesimal action of $\mathbf{gl}(E)$ so that $\alpha = D''f$ for some $f \in A^0(End(E))$, then $\hat{\varphi}(\alpha) = 0$. Since $tr(\alpha) = tr(D''f) = d''tr(f)$, we have

$$\hat{\varphi}(\alpha) = -\int_M d(\varphi \cdot tr(f) \wedge \Phi^{n-1}) = 0.$$

We show (ii) that $\hat{\varphi}$ is invariant under the action of $GL(E)$, If $f \in GL(E)$, then f sends $\alpha \in A^{0,1}(End(E)) \cong T_{D''}(\mathcal{D}''(E))$ to $f^{-1}\alpha f \in A^{0,1}(End(E)) \cong T_{D''f}(\mathcal{D}''(E))$. Since $tr(f^{-1}\alpha f) = tr(\alpha)$, it follows that

$$\hat{\varphi}(f^{-1}\alpha f) = \hat{\varphi}(\alpha).$$

This construction generalizes to a holomorphic p-form.

Each holomorphic p-form φ on M induces,in a natural way, a closed holomorphic p-form $\tilde{\varphi}$ on $\tilde{\mathcal{M}}(E)$.

For example, if φ is a holomorphic 2-form on M, then we set

$$\hat{\varphi}(\alpha, \beta) = \int_M \varphi \wedge tr(\alpha \wedge \beta) \wedge \Phi^{n-2}$$

for $\alpha, \beta \in A^{0,1}(End(E)) \cong T_{D''}(\mathcal{D}''(E))$.

If φ is a p-form, then in order to define $\hat{\varphi}(\alpha_1, \cdots, \alpha_p)$ for $\alpha_1, \cdots, \alpha_p \in A^{0,1}(End(E)) \cong T_{D''}(\mathcal{D}''(E))$, we need to skew-symmetrize $tr(\alpha_1 \wedge \cdots \wedge \alpha_p)$ in the definition.

If we use $tr(\alpha_1) \wedge \cdots \wedge tr(\alpha_p)$ in place of $tr(\alpha_1 \wedge \cdots \wedge \alpha^p)$, we would get only a p-form arising from the fibering $\mathcal{M}(E) \to Pic^0(M)$.

3. On Curvature of Moduli Spaces of Bundles over Curves. [1,3,5,8,9]

Let M be a compact Riemann surface of genus g, and let E be a C^∞ complex vector bundle of rank r over M. It follows from Theorem 2.1 that the moduli space $\tilde{\mathcal{M}}(E)$ of simple holomorphic structures on E is a (possibly non-Hausdorff) nonsingular complex manifold of dimension $r^2(g-1)+1$.

For a fixed line bundle $\mathcal{L}$, the moduli space $\tilde{\mathcal{M}}(E, \mathcal{L})$ of holomorphic structures with prescribed determinant bundle $\mathcal{L}$ has dimension $r^2(g-1)+1-g = (r^2-1)(g-1)$.

Since $D'' \circ D'' = 0$ automatically when $dim M = 1$, we have $\mathcal{D}''(E) = \mathcal{H}''(E)$. It follows that $\tilde{\mathcal{H}}''(E)$ and hence $\tilde{\mathcal{M}}(E)$ are connected.

Fix an Hermitian structure h in E. Then the moduli space $\tilde{\mathcal{M}}(E, h)$ of irreducible Einstein-Hermitian connections in E (or equivalently, the moduli space of stable holomorphic structures in E) is Hausdorff and open in $\tilde{\mathcal{M}}(E)$. Its natural Kähler metric has nonnegative holomorphic sectional curvature and hence nonnegative scalar curvature. If the holomorphic sectional curvature is identically zero, then $\mathcal{H}^{\mathcal{E}}(E, h)$ is totally geodesic (and hence flat) in $\mathcal{H}(E, h) = \mathcal{D}(E, h)$, and it follows that either M is an elliptic curve or the rank of E is 1. Except in these trivial cases, the scalar curvature of $\mathcal{M}(E, h)$ is strictly positive somewhere. It follows that except in the trivial cases where $\tilde{\mathcal{M}}(E, h)$ is a complex torus, the pluricanonical genera of $\tilde{\mathcal{M}}(E, h)$ are all zero provided $\tilde{\mathcal{M}}(E, h)$ is compact.

In the decomposition

$$H^1(M, \mathcal{O}(End(\mathcal{E}))) \cong H^1(M, \mathcal{O}(End^0(\mathcal{E}))) + H^1(M, \mathcal{O})$$

of the tangent space at $D'' \in \mathcal{H}^{\mathcal{E}}(E, h)$, the holomorphic sectional curvature vanishes in the direction of $H^1(M, \mathcal{O})$. The question remains whether it is positive in the direction of $H^1(M, \mathcal{O}(End^0(E^{D''})))$.

Although we can compute the curvature of $\tilde{\mathcal{M}}(E, h)$ explicitly using the Gauss equation for CR-submersions [8], the fact that the holomorphic sectional curvature is nonnegative can be understood intuitively from the general principle that the holomorphic curvature decreases with a holomorphic subbundle and increases with a quotient bundle. Thus, being a holomorphic quotient of a flat space $\mathcal{H}''(E) = \mathcal{D}''(E)$, the moduli space $\mathcal{M}(E)$ should have nonnegative holomorphic sectional curvature. However, if $dim M \geq 2$, then $\mathcal{H}''(E)$ is a complex submanifold of $\mathcal{D}''(E)$ and hence has nonpositive holomorphic sectional curvature. So passing to the quotient $\mathcal{M}(E)$, we should get both negative and positive terms in the expression for its holomorphic sectional curvature. In other words, when

$dimM \geq 2$, all we can say is that the holomorphic sectional curvature of $\mathcal{M}(E)$ is no more negative than that of $\mathcal{H}''(E)$.

4. Holomorphic Symplectic Structures. [9,12]

We consider now Kähler manifolds with holomorphic symplectic structure or hyper-Kähler manifolds. While Atiyah's article in this volume treats hyper-Kähler manifolds more from the real or quaternionic viewpoint, we shall emphasize the complex analytic viewpoint. For the quaternionic apporach to Theorems 4.1 and 4.2, the reader is referred to Atiyah's article and references therein.

A hyper-Kähler manifold is a Kähler manifold with a nongenerate holomorphic, parallel 2-form. In terms of holonomy, it is a Kähler manifold of even dimension, say $2m$, whose holonomy group is contained in $Sp(m)$. Such a Kähler manifold is automatically Ricci-flat.

The classical de Rham decomposition of a Riemannian manifold into irreducible factors takes a much more precise form in the special case of a Ricci-flat compact Kähler manifold. In fact,[7]:

If X is a compact Ricci-flat Kähler manifold, then it has a finite, unramified covering

$$\tilde{X} = T \times M_1 \times \cdots \times M_j \times N_1 \times \cdots \times N_k,$$

where T is a flat, complex torus, the M_i are compact, simply connected irreducible hyper-Kähler manifolds , and the N_i are compact, simply connected irreducible Kähler manifolds which have no holomorphic forms except for the form trivializing the canonical line bundle.

In terms of holonomy, the M_i have holonomy $Sp(m_i)$,(where $2m_i = dimM_i$), and the N_i have holonomy $SU(n_i)$, (where $n_i = dimN_i$), while T has trivial holonomy.

Thirty years ago, Wakakuwa [16] wrote a paper on topology of hyper-Kähler manifolds. Among other things, he proved that all odd dimensional Betti numbers are divisible by 4. However, for lack of interesting examples, Wakakuwa's paper has been long forgotten. Ten years ago, Bogomolov announced that there are no irreducible compact hyper-Kähler manifolds in dimension greater than 4. However, the first examples of 4-dimensional compact irreducible hyper-Kähler manifolds were discovered by Fujiki in 1981. He started with a K3 surface S and showed that a holomorphic (symplectic) 2-form of S induces a holomorphic symplectic form on the 4-dimensional complex manifold obtained by blowing up the 2-fold symmetric product $S^{(2)}$ along its diagonal,see [4]. His construction has been further generalized by Beauville to yield other examples,(see [2],[13]). However, there are still relatively few examples of compact hyper-Kähler manifolds. It seems natural to consider more generally holomorphic 2-forms which may be degenerate along proper subvarieties. These generalized holomorphic symplectic manifolds occur as non-singular models of k-fold symmetric products $X^{(k)}$ of holomorphic symplectic manifolds X. They occur also as moduli spaces of simple or stable holomorphic vector bundles over a compact holomorphic symplectic manifolds as we see in this section.

In order to discuss such moduli spaces, we consider infinite dimensional holomorphic symplectic manifolds. Let V be a complex Banach manifold. A holomorphic symplectic structure on V is given by a holomorphic 2-form ω_V which is closed and non-degenerate in the following sense:

$$\omega_V(X,Y) = 0 \quad \forall Y \in T_xV \quad \Rightarrow X = 0.$$

This condition means that the linear map $T_xV \to T_x^*V$ defined by ω_V is injective (but not necessarily bijective).

Let G be a complex Banach Lie group acting on V as a group of holomorphic transformations preserving the symplectic form ω_V. Let $\mathbf{g}$ be its Lie algebra and $\mathbf{g}^*$ the dual (Banach) space of $\mathbf{g}$. A holomorphic momentum map for the action of G on V is a holomorphic map $\mu : V \to g^*$ such that

$$< A, d\mu(X) >= \omega(A_x, X) \qquad \text{for} \quad A \in \mathbf{g}, X \in T_xV,$$

where $A_x \in T_xV$ is the vector defined by A through the action of G and $d\mu : T_xV \to \mathbf{g}^*$ is the differential of μ at $x \in V$.

We impose the following three conditions $(a), (b), (c)$ on μ.

(a) Assume that μ is G-equivariant in the sense that

$$\mu(g(x)) = (adg)^*(\mu(x)) \qquad \text{for} \quad g \in G, x \in V,$$

where $g \mapsto (adg)^*$ is the coadjoint representation of G.

(b) Assume that $0 \in \mathbf{g}^*$ is a weakly regular value of μ in the sense that

(b.1) $\mu^{-1}(0)$ is a complex submanifold of $V, (j : \mu^{-1}(0) \to V$ denoting the imbedding),

(b.2) the inclusion $dj : T_x(\mu^{-1}(0)) \subset Ker(d\mu_x)$ is an equality for every $x \in \mu^{-1}(0)$.

(c) Assume that the action of G is free on $\mu^{-1}(0)$ and that at each point $x \in \mu^{-1}(0)$ there is a holomorphic slice $S_x \subset \mu^{-1}(0)$ for the action, i.e., a complex submanifold S_x through x which is transversal to the orbit $G(x)$ in the sense that

$$T_x(\mu^{-1}(0)) = T_x(S_x) \oplus T_x(G(x)).$$

The quotient space $W = \mu^{-1}(0)/G$ is called the reduced phase space. If we take S_x sufficiently small, then the projection $\pi : \mu^{-1}(0) \to W$ defines a homeomorphism of S_x onto an open subset $\pi(S_x)$ of W and introduces a local coordinate system in W. This makes W into a (not necessarily Hausdorff) complex manifold. If the action of G on $\mu^{-1}(0)$ is proper, then W is Hausdorff.

We state now the holomorphic version of the symplectic reduction theorem of Marsden-Weinstein.

Theorem (4.1). *Let V be a complex Banach manifold with a holomorphic symplectic form ω_V. Let G be a complex Banach Lie group acting on V leaving ω_V invariant. If there is a holomorphic momentum map μ satisfying (a),(b) and (c), there exists a unique holomorphic symplectic form ω_W on the reduced phase space $W = \mu^{-1}(0)/G$ such that*

$$\pi^*\omega_W = j^*\omega_V \qquad on \quad \mu^{-1}(0).$$

We apply the reduction theorem to prove

Theorem (4.2). *Let M be a compact complex manifold with a holomorphic symplectic form ω_M and E a C^∞ complex vector bundle over M. Then ω_M induces, in a natural way, a holomorphic symplectic form on the nonsingular part of $\tilde{\mathcal{M}}(E)$.*

This theorem was first proved by Mukai [12] when M is an abelian surface or a K3 surface.

For the proof of Theorem 4.2, we apply the reduction theorem to

$$V = L^2_k(\mathcal{D}''(E)), \quad G = L^2_{k+1}(GL(E))/\mathbf{C}^*, \quad \mathbf{g} = L^2_{k+1}(\mathbf{gl}(E))/\mathbf{C},$$

where L^2_k and L^2_{k+1}, $(k > dimM)$, denote the Sobolev completion.

We define a holomorphic symplectic form ω_V on V by

$$\omega_V(\alpha, \beta) = \int_M tr(\alpha \wedge \beta) \wedge \omega_M^m \wedge \bar{\omega}_M^{m-1} \qquad \alpha, \beta \in T_{D''}(V),$$

where α and β are considered as elements of $L^2_k(A^{0,1}(End(E))) \cong T_{D''}(V)$ and $2m$ denotes the complex dimension of M.

We define a holomorphic momentum map $\mu : V \to \mathbf{g}^*$ by

$$< a, \mu(D'') >= -\int_M tr(a \circ D'' \circ D'') \wedge \omega_M^m \wedge \bar{\omega}_M^{m-1}, \qquad a \in \mathbf{g}, D'' \in V.$$

We note that

$$\mu^{-1}(0) = \{D'' \in V; D'' \circ D'' = 0\} = L^2_k(\mathcal{H}''(E)).$$

Remark. We showed in Section 2 that every holomorphic p-form φ on a compact Kähler manifold M induces a holomorphic p-form $\tilde{\varphi}$ on the moduli space $\tilde{\mathcal{M}}(E)$. We notice that the construction of $\tilde{\varphi}$ in Section 2 is quite similar to that of ω_V above. In fact, the only difference is that $(2m-1)$-th power Φ^{2m-1} of the Kähler form Φ is used in Section 2 in place of $\omega_M^{m-1} \wedge \bar{\omega}_M^{m-1}$ above. It is not hard to check that these two constructions give the same holomorphic 2-form (up to a constant factor) on $\tilde{\mathcal{M}}(E)$ when M is hyper-Kähler. The construction in Section 2 shows that if ω_M is a generalized holomorphic symplectic form, we still get a generalized holomorphic symplectic form on $\tilde{\mathcal{M}}(E)$.

5. Examples.

Let M be a complex torus of dimension n. The holomorphic vector bundles over an elliptic curve have been completely determined by Atiyah [1]. We consider therefore the two-dimensional case.

In general, if M is a compact Kähler surface with nowhere vanishing holomorphic 2-form ω_M, (i.e., if M is a 2-dimensional complex torus or a K3 surface), and if E is a C^∞ complex vector bundle of rank r over M, then by Serre duality $H^2(M, \mathcal{O}(End^0(E^{D''})))$ is dual to $H^0(M, \mathcal{O}(End^0(E^{D''})))$ for any $D'' \in \mathcal{H}''(E)$. Hence, at $[D''] \in \tilde{\mathcal{M}}(E)$ not only $H^0(M, \mathcal{O}(End^0(E^{D''})))$ but also $H^2(M, \mathcal{O}(End^0(E^{D''})))$ vanishes. So in this case the moduli space $\tilde{\mathcal{M}}(E)$ of simple holomorphic structures in E is nonsingular. From the Riemann-Roch formula we obtain its dimension:

$$dimH^1(M, \mathcal{O}(End(E^{D''}))) = \lambda(E) + r^2 h^{0,1} + 2 - 2r^2$$

where

$$\lambda(E) = 2rc_2(E) - (r-1)c_1(E)^2,$$

provided that $\tilde{\mathcal{M}}(E)$ is nonempty. By Theorem 4.2, $\tilde{\mathcal{M}}(E)$ is holomorphic symplectic. In particular, the dimension above is even, i.e., $(r-1)c_1(E)^2$ must be even if $\tilde{\mathcal{M}}(E)$ is nonempty.

Now we assume that M is a complex 2-dimensional torus. Then the moduli space $\tilde{\mathcal{M}}(E)$ has dimension

$$\lambda(E) + 2,$$

again provided that it is nonempty.

Assuming that $E^{D''}$ is semistable for some $D'' \in \mathcal{H}''(E)$, we have (Bogomolov's inequality)

$$\lambda(E) \geq 0.$$

If the equality holds here, the moduli space $\tilde{\mathcal{M}}(E)$ is a surface with a holomorphic symplectic form. We claim that this surface is a torus. In fact, as we saw in Section 2, the action of $Pic^0(M)$ alone gives a 2-dimensional family of holomorphic structures. So, the component of $\tilde{\mathcal{M}}(E)$ containing $\mathcal{E} = E^{D''}$ is isomorphic to $Pic^0(M)/\Sigma^0(\mathcal{E})$.

In general, if M is a complex torus of dimension n, every element $x \in M$ defines a translation $\tau_x : M \to M$ which sends $y \in M$ to $x + y \in M$. Given a holomorphic vector bundle $\mathcal{E}$ over M, every translation $\tau_x, x \in M$ induces a holomorphic vector bundle $\tau_x^*\mathcal{E}$. Thus, M acts on the moduli space $\tilde{\mathcal{M}}(E)$ of simple holomorphic structures in E.

We know from Section 2 that $Pic^0(M)$ acts on $\tilde{\mathcal{M}}(E)$ essentially effectively (for any compact Kähler manifold M). If M is a torus, then $M \times Pic^0(M)$ acts on $\tilde{\mathcal{M}}(E)$. Namely, every $(x, \mathcal{L}) \in M \times Pic^0(M)$ maps

$$(x, \mathcal{L}) : \mathcal{E} \mapsto \tau_x^*\mathcal{E} \cdot \mathcal{L}^{-1}.$$

Following Mukai [11], we set

$$\Phi^0(\mathcal{E}) = \{(x, \mathcal{L}) \in M \times Pic^0(M); \tau_x^*\mathcal{E} \cong \mathcal{E} \cdot \mathcal{L}\}.$$

Since $\Phi^0(\mathcal{E})$ is the isotropy subgroup of $M \times Pic^0(M)$ at $\mathcal{E} \in \tilde{\mathcal{M}}(E)$, the quotient group $(M \times Pic^0(M))/\Phi^0(\mathcal{E})$ embeds into $\tilde{\mathcal{M}}(E)$ as the $(M \times Pic^0(M))$-orbit through $\mathcal{E}$. In particular, we have

$$n \leq 2n - dim\Phi^0(\mathcal{E}) \leq dimH^1(M, \mathcal{O}(End(\mathcal{E}))).$$

The bundle $\mathcal{E}$ is said to be *semihomogeneous* if for every $x \in M$ there is a holomorphic line bundle $\mathcal{L}$ such that

$$\tau_x^*\mathcal{E} = \mathcal{E} \cdot \mathcal{L}.$$

Clearly, $\mathcal{E}$ is semihomogeneous if and only if $dim\Phi^0(\mathcal{E}) = n$. From the inequalities above it follows that if $dimH^1(M, \mathcal{O}(End(\mathcal{E}))) = n$, then $\mathcal{E}$ is semihomogeneous. An extensive study of semihomogeneous vector bundles has been made by Mukai. One of his main theorem states [11]

Theorem (5.1). *For a simple holomorphic vector bundle $\mathcal{E}$ on an abelian variety M, of dimension n, the following conditions are mutually equivalent.*

(1) $dimH^1(M, \mathcal{O}(End(\mathcal{E}))) = n$;

(2) $dimH^i(M, \mathcal{O}(End(\mathcal{E}))) = \binom{n}{i}$ for $i = 1, \cdots, n$;

(3) $\mathcal{E}$ is semihomogeneous;

(4) $End(\mathcal{E})$ is homogeneous, i.e., $\tau_x^*End(\mathcal{E}) \cong End(\mathcal{E})$ for all $x \in M$;

(5) there exist an isogeney $\pi : N \to M$ and a line bundle $\mathcal{L}$ over N such that $\mathcal{E} \cong \pi_*\mathcal{L}$.

We return to the two-dimensional case. The case of a vector bundle E over a 2-dimensional torus with $\lambda(E) = 0$ discussed above is the extreme case where $dim\Phi^0(\mathcal{E}) = 2 = n$.

The other extreme is reached by a Picard bundle. Let C be a nonsingular curve of genus 2, and J its Jacobian variety. Let $C^{(m)}$ denote the m-fold symmetric product of C. Fixing a point $x_0 \in C$, we define a map $\varphi : C^{(m)} \to J$ by

$$\varphi(x_1, \cdots, x_m) = x_1 + \cdots + x_m - mx_0 \qquad \text{for} \quad x_1, \cdots, x_m \in C.$$

For $m \geq 3, \varphi : C^{(m)} \to J$ is a projective bundle over J, and by Umemura [14] there is a stable vector bundle $\mathcal{E}$ of rank $m-1$ over J with $\lambda(\mathcal{E}) = 2$ such that $C^{(m)} = \mathbf{P}(\mathcal{E})$. Hence, $dimH^1(M, \mathcal{O}(End(\mathcal{E}))) = 4$. Let E denote the C^∞ complex vector bundle underlying $\mathcal{E}$. According to Umemura [15], the moduli space $\tilde{\mathcal{M}}(E)$ is isomorphic to $J \times Pic^0(J)$. So this is the case where $\Phi^0(\mathcal{E})$ is trivial.

We consider another example also due to Umemura [15]. Let A be an abelian surface with a prinicipal polarization $\mathcal{L}$, i.e., an ample line bundle with $dimH^0(A, \mathcal{L}) = 1$. (By the Riemann- Roch formula, this latter condition is equivalent to $c_1(\mathcal{L})^2 = 2$). Let $r \geq 2$, and let $\mathcal{L}_0, \mathcal{L}_1, \cdots, \mathcal{L}_r$ be line bundles algebraically equivalent to $\mathcal{L}$ with nonzero holomorphic sections $\varphi_0, \varphi_1, \cdots, \varphi_r$. Let $\mathcal{E}(\mathcal{L}_0, \mathcal{L}_1, \cdots, \mathcal{L}_r)$ be the coherent sheaf over A defined by the exact sequence:

$$0 \to \mathcal{O} \to \mathcal{L}_0 \oplus \mathcal{L}_1 \oplus \cdots \oplus \mathcal{L}_r \to \mathcal{E}(\mathcal{L}_0, \mathcal{L}_1, \cdots, \mathcal{L}_r) \to 0,$$

where the second arrow is given by

$$1 \mapsto (\varphi_0, \varphi_1, \cdots, \varphi_r).$$

The isomorphism class of $\mathcal{E}(\mathcal{L}_0, \mathcal{L}_1, \cdots, \mathcal{L}_r)$ does not depend on the choice of $\varphi_0, \varphi_1, \cdots, \varphi_r$.
From the Whitney sum formula and $c_1(\mathcal{L}_i)^2 = 2$ we obtain

$$\lambda(\mathcal{E}(\mathcal{L}_0, \mathcal{L}_1, \cdots, \mathcal{L}_r)) = 2(r+1).$$

Hence, we expect that the dimension of the moduli space of $\mathcal{E}(\mathcal{L}_0, \mathcal{L}_1, \cdots, \mathcal{L}_r)$ to be $2(r+2)$.
Using the translations $\tau_{x_0}, \tau_{x_1}, \cdots, \tau_{x_r}$ by elements $x_0, x_1, \cdots, x_r \in A$, we set

$$\mathcal{E}_{x_0, x_1, \cdots, x_r} = \mathcal{E}(\tau_{x_0}^* \mathcal{L}, \tau_{x_1}^* \mathcal{L}, \cdots, \tau_{x_r}^* \mathcal{L}).$$

Then, for generic $x_0, x_1, \cdots, x_r \in A, \mathcal{E}_{x_0, x_1, \cdots, x_r}$ is an $\mathcal{L}$-stable vector bundle of rank r, and under the map

$$(\mathcal{L}, x_0, x_1, \cdots, x_r) \mapsto \mathcal{E}_{x_0, x_1, \cdots, x_r}$$

the algebraic variety $Pic^0(A) \times A^{(r+1)}$ is birationally isomorphic to a component of the moduli space of $\mathcal{L}$-stable vector bundle containing a generic $\mathcal{E}_{x_0, x_1, \cdots, x_r}$. (Here, $A^{(r+1)}$ denotes the $(r+1)$-fold symmetric product of A).

Remark. If E is a C^∞ complex vector bundle of rank r over a 4-dimensional complex torus M, we can still express the dimension of its moduli space in terms of Chern classes $c_i = c_i(E)$ of E. Let $D'' \in \tilde{\mathcal{H}}''(E)$ and $\mathcal{E} = E^{D''}$. Since $\mathcal{E}$ is simple, we have $dimH^0(M, \mathcal{O}(End(\mathcal{E}))) = 1$. At a nonsingular point $[D'']$, we have $H^2(M, \mathcal{O}(End^0(\mathcal{E}))) =$

0 so that $dim H^2(M, \mathcal{O}(End(\mathcal{E}))) = 6$. Hence, by the Riemann-Roch formula and Serre duality we obtain

$$dim H^1(M, \mathcal{O}(End(\mathcal{E}))) = 4 + \nu(E),$$

where

$$\nu(E) = -\frac{1}{2}\int_M ch(End(E)),$$

or, in terms of $c_i = c_i(E)$,

$$\nu(E) = \frac{1}{24}\{-(r-1)c_1^4 + 4rc_1^2c_2 - 2(r+6)c_2^2 - 4(r-3)c_1c_3 + 4rc_4\}.$$

We do not know if $\nu(E)$ is nonnegative for a semistable E over any compact 4-dimensional Kähler manifold.

Mukai [12] has given some interesting examples of stable vector bundles over a K3 surface. For example, let M be a nonsingular complete intersection of three hyperquadrics in $\mathbf{P}_5$ such that all hyperquadrics containing M are of rank 5. Let $h \in H^2(M, \mathbf{Z})$ be the cohomology class given by a hyperplane section. Then the moduli space of h-stable vector bundles $\mathcal{E}$ of rank 2 over M with $c_1(\mathcal{E}) = h$ and $c_2(\mathcal{E}) = 4$ is a K3 surface that can be described as a two-fold covering of $\mathbf{P}_2$ ramified over a curve of degree 6.

In another example of Mukai, M is a nonsingular complete intersection of a hyperquadric and a hypercubic in $\mathbf{P}_4$. Assume that M contains no lines. Then the moduli space of h-stable vector bundles $\mathcal{E}$ of rank 2 over M with $c_1(\mathcal{E}) = -h$ and $c_2(\mathcal{E}) = 4$ is a symplectic Kähler manifold of dimension 4, $Hilb^2M$, discovered by Fujiki [4].

As we stated at the bigining of Section 4, every compact hyper-Kähler manifold X has a finite, unramified covering $\tilde{X}$ which factors into a product of (even-dimensional) complex torus and a simply connected hyper-Kähler manifold. The above examples seem to indicate that, for the moduli space $\tilde{\mathcal{M}}(E)$ or $\tilde{\mathcal{M}}(E, h)$, its torus factor comes only from the torus factor of M.

References

[1]. M.F. Atiyah, Vector bundles over an elliptic curves, Proc. London Math. Soc. (3) 7 (1957), 414-452.

[2]. A. Beauville, Variétés kählériennes dont la première classe de Chern est nulle, J. of Diff. Geometry 18 (1983), 755-782.

[3]. K. Cho, Positivity of the curvature of the Weil-Petersson metric on the moduli space of stable vector bundles, Ph. D. thesis, Harvard Univ., 1985.

[4]. A. Fujiki, On primitively symplectic compact Kähler V-manifolds of dimension four, in *Classification of Algebraic and Analytic Manifolds*, pp. 71-250, Progr. Math. 39, Birkhäuser, 1983

[5]. M. Itoh, Geometry of anti-self-dual connections and Kuranishi map, J. Math. Soc. Japan 40 (1988), 9-33.

[6]. H. J. Kim, Moduli of Hermite-Einstein vector bundles, Math. Z. 195 (1987), 143-150.

[7]. S. Kobayashi, Recent results in complex differential geometry, Jber. d. Dt. Math.-Verein. 83 (1981), 147-158.

[8]. S. Kobayashi, Submersions of CR submanifolds, Tohoku Math. J. 39 (1987), 95-100.

[9]. S. Kobayashi, *Differential Geometry of Complex Vector Bundles*, Iwanami Shoten/ Princeton U. Press, 1987.

[10]. M. Lübke and C. Okonek, Moduli spaces of simple bundles and Hermitian-Einstein connections, Math. Ann. 276 (1987), 663-674.

[11]. S. Mukai, Semi-homogeneous vector bundles on an abelian variety, J. Math. Kyoto Univ. 18 (1978), 239-272.

[12]. S. Mukai, Symplectic structure of the moduli space of sheaves on an abelian or K3 surface, Invent. Math. 77 (1984), 101-116.

[13]. A. J. Smith, Symplectic Kähler manifolds, Ph. D. thesis, Univ. of Calif., Berkeley, 1987.

[14]. H. Umemura, On a property of symmetric products of a curve of genus 2, Proc. Intl. Symp. on Algebraic Geometry, Kyoto 1977, pp.709-721.

[15]. H. Umemura, Moduli spaces of the stable vector bundles over abelian surfaces, Nagoya Math. J. 77 (1980), 47-60.

[16]. H. Wakakuwa, On Riemannian manifolds with homogeneous holonomy group $Sp(n)$, Tohoku Math. J. 10 (1958), 274-303.

This work has been partially supported by NSF Grant DMS-8801371.

Quasiconformal Mappings on CR Manifolds

Adam Korányi[1]
Hans Martin Reimann[2]

1 CR Manifolds and the Definition of Quasiconformal Mapping

A nondegenerate differential one-form θ on a $(2n+1)$-dimensional manifold M defines a contact structure on M. The contact structure is the subbundle HM of the tangent bundle TM consisting of the horizontal vectors:

$$HM = \{X \in TM : \theta(X) = 0\}$$

The form θ is only determined up to a nonvanishing scalar factor and there is a nondegeneracy condition stating that $\theta \wedge (d\theta)^n$ is a nonvanishing volume form. It is sufficient to have θ defined locally. But it will be assumed that M is orientable which implies that θ can be defined globally.

If in addition to the contact structure there is given a smoothly varying automorphism $J: HM \to HM$ with the property that $J^2 = -I$ (i.e. a "complex structure"), then M has the structure of a CR manifold.

The complexified subbundle $H_{\mathbf{C}}M$ decomposes into the eigenspaces for J with eigenvalues $+i$ and $-i$

$$H_{\mathbf{C}}M = T^{1,0} + T^{0,1}, \qquad T^{1,0} \cap T^{0,1} = \{0\}$$

$T^{1,0}$ is the space of holomorphic and $T^{0,1} = \overline{T^{1,0}}$ the space of antiholomorphic vectors.

Locally, the space $T^{1,0}$ can be defined by complex one-forms $\theta^\alpha \in (T^{1,0})^*$, $\alpha = 1, \ldots, n$ such that together with their complex conjugates $\theta^{\overline{\alpha}} = \overline{\theta^\alpha}$ and with θ they form a basis of the complexified cotangent space $T^*_{\mathbf{C}}M$. The CR manifold M is called integrable if

$$d\theta \equiv d\theta^\alpha \equiv 0 \pmod{\theta, \theta^1, \ldots, \theta^n}$$

for all α. We will assume throughout that this integrability condition is satisfied. For a fixed choice of the contact form θ the forms θ^α can then be determined such that $d\theta = i g_{\alpha\overline{\beta}} \theta^\alpha \wedge \theta^{\overline{\beta}}$ (Webster [15]) with $(g_{\alpha\overline{\beta}})$ a Hermitian matrix. In particular $d\theta(Z, W) = 0$ for $z, w \in T^{1,0}$, and $d\theta$ is J-invariant

$$d\theta(JX, JY) = d\theta(X, Y) \text{ for } X, Y \in H_{\mathbf{C}}M.$$

The Levi form L is the Hermitian form on $H_{\mathbf{C}}M$ which is uniquely determined by

$$L(Z, Z) = -i d\theta(Z, \overline{Z}) \text{ for } Z \in T^{1,0}$$

[1]Lehman College and Graduate Center, City University of New York

[2]University of Bern, Switzerland

For real vectors X, Y the Levi form can expressed as

$$L(X,Y) = d\theta(X, JY) = d\theta(Y, JX).$$

If Z_α, $\alpha = 1, \dots, n$ is the basis of $T^{1,0}$ dual to θ^α, $\alpha = 1, \dots, n$ then $L(Z_\alpha, Z_\beta) = g_{\alpha\overline{\beta}}$. The Levi form depends on the choice of the contact form θ. However if θ and $\lambda\theta$ are contact forms defining the same contact structure, then the associated Levi forms are the same up to the scalar factor λ.

The CR manifold M is strictly pseudoconvex if the Levi form is positive on HM. Replacing θ by $\lambda\theta$ on a strictly pseudoconvex CR manifold then requires that λ be positive. Note also that the nondegeneracy condition $\theta \wedge (d\theta)^n \neq 0$, which usually is not part of the definition of a CR manifold, is automatically satisfied on strictly pseudoconvex CR manifolds.

The smooth quasiconformal mappings $f: M \to M'$ between strictly pseudoconvex integrable CR manifolds will be defined as C^2-contact diffeomorphisms which are of bounded distortion. If θ and θ' are contact forms on M and M' with associated Levi forms L and L' then the diffeomorphism $f: M \to M'$ is a contact transformation if $f^*\theta' = \lambda\theta$ with some nonvanishing scalar function λ. We will restrict all considerations to the case $\lambda > 0$.

DEFINITION *A smooth K-quasiconformal mapping $f: M \to M'$ between strictly pseudoconvex CR manifolds is a C^2-contact transformation such that $f^*\theta' = \lambda\theta$ with $\lambda > 0$ and*

$$\lambda K^{-1} L(X,X) \leq L'(f_*X, f_*X) \leq \lambda K L(X,X)$$

for all $X \in HM$.

It is proved in [10] that for smooth mappings of the Heisenberg group to itself, this definition is equivalent to the original definition of quasiconformality due to Mostow. The smoothness requirement can in fact be relaxed considerably. We will come back to this question in a subsequent paper.

The distortion condition can be formulated in an equivalent way which puts in evidence that quasiconformal mappings change the complex structure. Under the contact transformation $f: M \to M'$ the complex structure $J': HM' \to HM'$ pulls back to an automorphism

$$J^* = (f_*)^{-1} \circ J' \circ f_*$$

on HM, which again satisfies $(J^*)^2 = -I$. If L^* denotes the Levi form with respect to J_* and the contact form θ, then

$$L'(f_*X, f_*X) = d\theta'(f_*X, f_*(f_*^{-1}J'f_*X)) = \lambda d\theta(X, J^*X) = \lambda L^*(X,X).$$

Therefore, f is K-quasiconformal if and only if

$$K^{-1}L(X,X) \leq L^*(X,X) \leq KL(X,X)$$

for all $X \in HM$.

If f is quasiconformal and $0 \not\equiv \overline{Z} \in T^{0,1}$ then $f_*(\overline{Z})$ cannot be contained in $(T')^{1,0}$ because this would lead to $L'(f_*\overline{Z}, f_*\overline{Z}) = -id\theta'(f_*\overline{Z}, f_*Z) = -\lambda L(Z,Z)$, contradicting the positive definiteness of L. The subspace

$$T_*^{1,0} = \{Z \in H_{\mathbf{C}}M : f_*Z \in (T')^{1,0}\}$$

can therefore be represented as $T_*^{1,0} = \{Z - \overline{\mu Z} : Z \in T^{1,0}\}$ where $\mu: T^{1,0} \to T^{1,0}$ is a complex antilinear mapping. The J-invariance of $d\theta'$ implies the symmetry of μ:

$$\lambda d\theta(Z - \overline{\mu Z}, Z' - \overline{\mu Z'}) = d\theta'(f_*(Z - \overline{\mu Z}, Z' - \overline{\mu Z'})) = 0$$

for all $Z, Z' \in T^{1,0}$, and hence

$$d\theta(Z, \overline{\mu Z'}) - d\theta(Z', \overline{\mu Z}) = 0.$$

The mapping μ will be called the complex dilatation of the quasiconformal mapping. It is uniquely determined by the subspace $T_*^{1,0}$. In particular, if $f_j: M \to M^{(j)}$, $j = 1, 2$ are two quasiconformal mappings with the same complex dilatation μ, then the mapping $f_2 \circ f_1^{-1}$ is a quasiconformal mapping which preserves the complex structure, hence a conformal mapping. Its complex dilatation is 0. In the case that the CR manifolds are embedded as hypersurfaces in $\mathbf{C}^n$, the mapping $f_2 \circ f_1^{-1}$ locally extends to a biholomorphic mapping. (Webster [14]; see also [10] for the connection with the complex dilatation.)

As will be shown, a contact transformation $f: M \to M'$ is K-quasiconformal if and only if

$$\|\mu\| = \sup_{L(Z,Z)=1} L(\mu Z, \mu Z)^{1/2} \leq \frac{K-1}{K+1}.$$

Observe that $\|\mu\|$ only depends on the position of $T_*^{1,0}$ in the space $H_{\mathbf{C}}M$ of horizontal tangent vectors. For the proof of the statement, it therefore suffices to consider the special case where $M = M' = N$ is the Heisenberg group. The proof will be given in the next section.

2 The Beltrami Equation

Strongly pseudoconvex CR manifolds naturally arise as boundaries of the strongly pseudoconvex domains in $\mathbf{C}^{n+1}$. If ρ is a defining function for such a domain D

$$D = \{z \in \mathbf{C}^{n+1} : \rho(z) < 0\}$$

and $d\rho(z) \neq 0$ for $z \in \partial D$, then the differential $\theta = (\partial\rho - \overline{\partial\rho})/2i$ defines the contact structure on $M = \partial D$. The complex structure of the underlying space $\mathbf{C}^{n+1}$ induces the complex structure of HM.

The Heisenberg group $N = \partial D$

$$D = \{Z \in \mathbf{C}^{n+1} : \rho(Z) = \Im z_{n+1} - \sum_0^n |z_j|^2 > 0\}$$

serves as a model case. It is holomorphically equivalent via the Cayley transform to the unit sphere S with one point removed. The sphere S is distinguished by the property that the dimension of its (holomorphic) automorphism group is maximal (E. Cartan [2]).

In this section we consider quasiconformal mappings from N onto strictly pseudoconvex hypersurfaces in $\mathbf{C}^{n+1}$ and in particular from N onto N. We first give a different interpretation of the complex dilatation for such mappings $f = (f^1, \ldots, f^{n+1}): N \to \mathbf{C}^{n+1}$.

In particular, we show that f satisfies a system of differential equations which we call, in analogy to the one dimensional case of a quasiconformal mapping $f: \mathbb{C} \to \mathbb{C}$, a Beltrami equation.

For this purpose, introduce the basis

$$\begin{aligned} \theta &= \tfrac{1}{2i}\textstyle\sum_{\alpha=1}^{n} \overline{z_\alpha} dz_\alpha - \tfrac{1}{2i}\sum_{\alpha=1}^{n} z_\alpha d\overline{z_\alpha} + dt/2 \\ \theta^\alpha &= dz_\alpha \qquad\qquad\qquad\qquad\qquad\qquad \alpha = 1,\ldots,n \\ \theta^{\overline{\alpha}} &= d\overline{z_\alpha} \end{aligned}$$

in the space of one-forms on N (the points of N are parameterized by $(z,t) \in \mathbb{C}^n \times \mathbb{R}$, $t = \Re z_{n+1}$). The dual basis is then given by the vector fields $T = 2\frac{\partial}{\partial t}$, $Z_\alpha = \frac{\partial}{\partial t_\alpha} + i\overline{z_\alpha}\frac{\partial}{\partial t}$, $\overline{Z_\alpha}$, and $\{Z_1,\ldots,Z_n\}$ is an orthonormal basis of the space $T^{1,0}$ of holomorphic vectors with respect to the Levi form.

The image of Z_α under the tangent mapping f_* decomposes into a holomorphic and an antiholomorphic part: $f_* Z_\alpha = V_\alpha + \overline{W_\alpha}$, $V_\alpha, W_\alpha \in T^{1,0}$ with

$$V_\alpha = \sum_{k=1}^{n+1} (Z_\alpha f^k) \frac{\partial}{\partial z_k},$$

$$W_\alpha = \sum_{k=1}^{n+1} (\overline{Z_\alpha f^k}) \frac{\partial}{\partial \overline{z_k}}.$$

Since f is assumed to be a contact diffeomorphism between N and the strictly pseudoconvex hypersurface $M = f(N) \subset \mathbb{C}^{n+1}$, it satisfies (cf. the restriction on the type of the contact forms and the contact mappings introduced earlier) $f^*\theta' = \lambda\theta$ $(\lambda > 0)$ where θ' is a contact form on M. It then follows that the vectors $V_1,\ldots,V_n$ span the whole space of holomorphic tangent vectors. Otherwise there would exist $Z \in T^{1,0}$, $Z \neq 0$, with $f_*\, Z = \overline{W} \in T^{0,1}$ and this would contradict the strict pseudoconvexity:

$$0 < -\lambda i\theta(Z,\overline{Z}) = -i\theta'(\overline{W},W) \leq 0.$$

There exists therefore a complex antilinear mapping $\mu: T^{1,0} \to T^{1,0}$ with $W_\alpha = \sum_{\beta=1}^{n} \mu_{\alpha\beta} V_\beta$, $(\alpha = 1,\ldots,n)$. Observe now that the vectors $Z_\alpha - \sum \overline{\mu_{\alpha,\beta} Z_\beta}$ $(\alpha = 1,\ldots,n)$ are mapped into $T^{1,0}$:

$$\begin{aligned} &f_*(Z_\alpha - \textstyle\sum \overline{\mu_{\alpha,\beta} Z_\beta}) \\ &= V_\alpha + \textstyle\sum_\beta \overline{\mu_{\alpha,\beta}\, V_\beta} - \sum_\beta \overline{\mu_{\alpha,\beta}} \left(\overline{V_\beta} + \sum_\gamma \mu_{\beta,\gamma} V_\gamma\right) \\ &= V_\alpha - \textstyle\sum_{\beta,\gamma} \overline{\mu_{\alpha,\beta}} \mu_{\beta,\gamma} V_\gamma \in T^{1,0} \end{aligned}$$

The mapping μ therefore coincides with the complex dilatation μ introduced in the first section.

In view of the definition for V_α and W_α, the equation $W_\alpha = \sum_{\beta=1}^{n} \mu_{\alpha,\beta} V_\beta$ can be written as a general Beltrami equation

$$\overline{Z_\alpha} f^k = \sum_\beta \mu_{\alpha,\beta} Z_\beta f^k, \qquad (k = 1,\ldots,n+1).$$

We have thus shown that any contact diffeomorphism $f: N \to M$ from N onto a strictly pseudoconvex hypersurface M satisfies such a Beltrami equation with μ symmetric. Later on we will show that if f if K-quasiconformal then in addition

$$\|\mu\|^2 = \sup_{0 \neq Z \in T^{1,0}} \frac{L(\mu Z, \mu Z)}{L(Z,Z)} \leq \left(\frac{K-1}{K+1}\right)^2.$$

Conversely, assume that the diffeomorphism $f: N \to M$ satisfies the Beltrami equation, with the matrix $(\mu_{\alpha,\beta})$ symmetric. Then f is a contact diffeomorphism. In fact if

$$V_\alpha = \sum_{k=1}^{n+1} (Z_\alpha f^k) \frac{\partial}{\partial z_k},$$

$$W_\alpha = \sum_{k=1}^{n+1} (\overline{Z_\alpha} f^k) \frac{\partial}{\partial z_k}$$

then $f_* Z_\alpha = V_\alpha + \overline{W_\alpha}$ and $f_* Z_{\overline{\alpha}} = \overline{V_\alpha} + W_\alpha$ span a J-invariant subspace, since $W_\alpha = \sum \mu_{\alpha,\beta} V_\beta$. Hence there exists a holomorphic differential form τ on $\mathbb{C}^{n+1}$, $\tau \neq 0$, such that $\tau(V_\alpha) = 0$, $\alpha = 1, \ldots, n$.

Furthermore, since f_* is regular, a suitable linear combination θ' of $1/2(\tau + \overline{\tau})$ and $1/2(\tau - \overline{\tau})$ can be chosen such that $(f_* \tau)(\theta') = 1$. It then follows that $f^* \theta'|_N = \theta$.

If in addition it is assumed that $\|\mu\| < 1$, then $M = fN$ is strictly pseudoconvex, i.e. the Levi form L' associated to θ' is positive definite:

$$\begin{aligned}
&L'(f_*(Z - \overline{\mu Z}), f_*(Z - \overline{\mu Z})) \\
&= -id\theta'(f_*(Z - \overline{\mu Z}), \overline{f_*(Z - \overline{\mu Z})}) \\
&= -id\theta(Z, \overline{Z}) - id\theta(\overline{\mu Z}, \mu Z) \\
&= L(Z,Z) - L(\mu Z, \mu Z) \\
&\geq L(Z,Z)(1 - \|\mu\|) \\
&> 0.
\end{aligned}$$

As will be shown next, the condition $\|\mu\| \leq (K-1)(K+1)^{-1}$ then implies that f is K-quasiconformal.

THEOREM *The contact diffeomorphism $f: N \to M$ between strictly pseudoconvex integrable CR manifolds N and M is K-quasiconformal if and only if its complex dilatation μ satisfies*

$$\|\mu\| \leq \frac{K-1}{K+1}.$$

As already remarked in the first section, it suffices to consider a contact diffeomorphism of the Heisenberg group N onto itself.

With respect to the basis $\{e_1, \ldots, e_{2n}\}$,

$$\begin{aligned}
e_\alpha &= X_\alpha = \frac{\partial}{\partial x_\alpha} + 2y_\alpha \frac{\partial}{\partial t} \\
e_{\alpha+n} &= Y_\alpha = \frac{\partial}{\partial y_\alpha} + 2x_\alpha \frac{\partial}{\partial t}
\end{aligned}$$

the tangent mapping f_* restricted to HN is given by a multiple of a symplectic mapping $d\theta(f_*X, f_*Y) = \lambda d\theta(X,Y), \qquad (X,Y \in HN)$.

The mapping $(1/\lambda)f_*$ restricted to HN can thus be represented by an element g in $\mathrm{Sp}(2n,\mathbf{R}) = \{g \in \mathrm{Gl}(\mathbf{R}^{2n}) : g^{-1}Jg = J\}$, where $J = \begin{pmatrix} 0 & -I \\ I & 0 \end{pmatrix}$ (with I the identity matrix in $\mathbf{R}^n$).

By complexification g extends to a complex linear mapping of $\mathbf{C}^{2n}$. The subspace $T^{1,0} \subset \mathbf{C}^{2n}$ is spanned by the vectors $1/2(e_\alpha - ie_{n+\alpha}) = Z_\alpha$ $(\alpha = 1,\ldots,n)$. Upon setting $g(e_\alpha - ie_{n+\alpha}) = g(Z_\alpha) = V_\alpha + \overline{W_\alpha}$ with $V_\alpha, W_\alpha \in T^{1,0}$, the complex dilatation of g, $\mu = \mu(g)$, is defined by $W_\alpha = \sum_\beta \mu_{\alpha,\beta} V_\beta$.

We will now make use of the Cartan decomposition $G = KA^+K$ of $G = \mathrm{Sp}(2n,\mathbf{R})$. The subgroup $K = \{k \in G : k^t = k^{-1}\}$ is identified with $U(n)$. $T^{1,0}$ is an invariant subspace for $k \in K$ and $\kappa := k|_{T^{1,0}}$ is unitary on $T^{1,0} \approx \mathbf{C}^n$. The subgroup A is of the form

$$A = \{a = \begin{pmatrix} e^{t_1} & & & & & \\ & \ddots & & & & \\ & & e^{t_n} & & & \\ & & & e^{-t_1} & & \\ & & & & \ddots & \\ & & & & & e^{-t_n} \end{pmatrix} : t_\alpha \in \mathbf{R}; \alpha = 1,\ldots,n\}$$

and the elements in A^+ satisfy $t_1 \geq t_2 \geq \ldots \geq t_n \geq 0$. From the definition of μ it follows that

$$\begin{aligned} \mu(a) &= \begin{pmatrix} \tanh t_1 & & \\ & \ddots & \\ & & \tanh t_n \end{pmatrix} \\ \mu(gk) &= \kappa^{-1}\mu(g)\overline{\kappa} \qquad\qquad a \in A; k \in K; g \in G \\ \mu(kg) &= \mu(g) \end{aligned}$$

and consequently for $g = k_1ak_2$, $\mu(g) = \mu(k_1ak_2) = k_2^{-1}\mu(a)\overline{k_2}$. (In particular, $\mu(g)$ is symmetric.)

The norm of $\mu(g)\colon T^{1,0} \to T^{1,0}$ (with $g = k_1ak_2$ $(k_1, k_2 \in K$, $a \in A^+)$ is then given by

$$\|\mu\| = \sup_{Z \in T^{1,0}; |Z|=1} |\mu(g)Z| = \max_\alpha \tanh t_\alpha = \tanh t_1.$$

The condition of quasiconformality is expressed by the distortion inequality

$$K^{-1}|x| \leq |gx| \leq K|x| \text{ for all } x \in \mathbf{R}^{2n}.$$

But since

$$\frac{\max_{|x|=1}|gx|}{\min_{|x|=1}|gx|} = \frac{e^{t_1}}{e^{-t_1}} = \frac{1+\tanh t_1}{1-\tanh t_1} = \frac{1+\|\mu\|}{1-\|\mu\|}$$

the distortion inequality is equivalent to

$$\|\mu\| \leq \frac{K+1}{K-1}.$$

This completes the proof of the theorem.

It would now be interesting to know which functions μ arise as complex dilatations of quasiconformal mappings. Let us change the point of view slightly. Assume that $f: N \to \mathbf{C}^{n+1}$ is a K-quasiconformal mapping of the Heisenberg group onto some strictly pseudoconvex hypersurface $M = fN \subset \mathbf{C}^{n+1}$. Denote the complex dilatation by μ and introduce a new complex structure J_* on N by $J_* = (f_*)^{-1} \circ J' \circ f_*$, where J' is the complex structure of the image domain (or equivalently the complex structure on $\mathbf{C}^{n+1}$) and f_* the tangent mapping restricted to HN. The complex structure J_* is completely determined by μ. The Heisenberg group N with the contact form

$$\theta = \frac{1}{2i}\sum_{\alpha} \overline{z_\alpha} dz_\alpha - \frac{1}{2i}\sum_{\alpha} z_\alpha d\overline{z_\alpha} + \frac{dt}{2}$$

and this new complex structure J_* is a CR manifold, denoted by N_μ. A basis for the holomorphic differentials is given by $\gamma^\alpha = dz_\alpha + \sum_\beta \mu_{\alpha,\beta} d\overline{z_\beta}$ $(\alpha = 1, \ldots, n)$ and the dual basis for the vector fields is

$$V_\alpha = (1 - \sum_\beta \mu_{\alpha,\beta}\overline{\mu_{\alpha,\beta}})^{-1}(Z_\alpha - \sum \overline{\mu_{\alpha,\beta} Z_\beta}).$$

If μ is the complex dilatation of a K-quasiconformal mapping, then μ is symmetric and satisfies

$$\|\mu\| \leq \frac{K-1}{K+1}.$$

But in addition it also has to satisfy the (Frobenius) integrability condition $d\theta \equiv d\gamma^\alpha \equiv 0$ $(\text{mod } \theta, \gamma^1, \ldots, \gamma^n)$ for all α. This can easily be verified by a direct calculation.

Assume now that μ satisfies the above conditions. The question whether a quasiconformal mapping $f: N \to \mathbf{C}^{n+1}$ with this complex dilatation exists is then equivalent to the question whether N_μ is embeddable into $\mathbf{C}^{n+1}$. The statement that the CR manifold N_μ is equivalent to N precisely means that there exists a quasiconformal mapping $f: N \to N$ with complex dilatation μ.

Both the embedding and the equivalence problem have been treated extensively in the literature. The embedding problem was first posed by Kohn [9] and subsequently solved to a large extent by Kuranishi [12]. According to Kuranishi's result, every smooth strictly pseudoconvex, integrable CR manifold of real dimension $2n+1 \geq 9$ arises locally as the boundary of a strictly pseudoconvex domain in $\mathbf{C}^{n+1}$. Akahori [1] extended this result to dimension 7. A greatly simplified proof has recently been given by Webster [16]. The problem remains open for dimension 5, whereas counterexamples in dimension 3 are due to Nirenberg [13]. These are constructed as perturbations of the CR structure of the Heisenberg group. More generally Jacobowitz and Trèves [8] have shown that analytically small perturbations of strictly pseudoconvex realizable CR structures of dimension 3 result in nonembeddable CR manifolds.

The equivalence problem was solved by E. Cartan [2] for dimension 3. He constructed a fifth order relative CR invariant whose vanishing is the only (necessary and sufficient) condition for equivalence of a given 3-dimensional CR structure with the standard structure of the Heisenberg group. Cartan's method also applies to the higher dimensional case. In [3] Chern and Moser derived the corresponding relative CR invariants and thus solved the local equivalence problem. The observation that for dimensions $2n+1 \geq 5$ the higher order invariants can all be reduced to the invariants of second order is due to Webster [15]. Let us briefly summarize these results for $2n+1 \geq 5$.

On the integrable, strictly pseudoconvex CR manifold M fix a contact form θ and a basis $\theta^1, \ldots, \theta^n$ of $T^{1,0}$ such that $d\theta = +ig_{\alpha\overline{\beta}}\theta^\alpha \wedge \theta^{\overline{\beta}}$ with $(g_{\alpha\overline{\beta}})$ Hermitian and positive definite. The connection forms $\omega_\beta{}^\alpha$ and torsion forms τ^α are the uniquely defined forms satisfying

$$\begin{aligned} d\theta^\alpha &= \theta^\beta \wedge \omega^\alpha_\beta + \theta \wedge \tau^\alpha \\ \tau^\alpha &\equiv 0 \pmod{\theta^{\overline{\beta}}} \end{aligned} \qquad \alpha = 1, \ldots, n$$

and the symmetry condition

$$dg_{\alpha\overline{\beta}} - \omega_\alpha{}^\gamma g_{\gamma\overline{\beta}} - \omega_{\overline{\beta}}{}^{\overline{\gamma}} g_{\alpha\overline{\gamma}} = 0$$

(the summation convention is being used). If the curvature forms are defined by

$$\Omega_\beta{}^\alpha = d\omega_\beta{}^\alpha - \omega_\beta{}^\alpha \wedge \omega_\gamma{}^\alpha - i\theta_\beta \wedge \tau^\alpha + i\tau_\beta \wedge \theta^\alpha$$

where the Levi form $g_{\alpha\overline{\beta}}$ and its inverse $g^{\overline{\beta}\alpha}$ are used to lower and raise indices, then

$$\Omega_{\beta\overline{\alpha}} = R_{\beta\overline{\alpha}\sigma\overline{\delta}}\theta^\sigma \wedge \theta^{\overline{\delta}} + \lambda_{\beta\overline{\alpha}} \wedge \theta$$

for some one-forms $\lambda_{\beta\overline{\alpha}}$ and curvature tensor components $R_{\beta\overline{\alpha}\sigma\overline{\delta}}$.

The curvature satisfies the following symmetry conditions:

$$\begin{aligned} R_{\beta\overline{\alpha}\sigma\overline{\delta}} &= R_{\overline{\alpha}\beta\overline{\delta}\sigma} \quad (:= \overline{R}_{\alpha\overline{\beta}\delta\overline{\sigma}}) \\ R_{\beta\overline{\alpha}\sigma\overline{\delta}} &= R_{\sigma\overline{\alpha}\beta\overline{\delta}}. \end{aligned}$$

The pseudoconformal curvature tensor S given by Chern and Moser [3] can then be expressed in terms of R (Webster [16, p. 35]),

$$\begin{aligned} S_{\beta\rho}{}^\alpha{}_{\overline{\sigma}} &= R_\beta{}^\alpha{}_{\rho\overline{\sigma}} \\ &= R_\beta{}^\alpha{}_{\rho\overline{\sigma}} - \frac{1}{n+2}(R_\beta{}^\alpha{}_{\rho\overline{\sigma}} + R_\rho{}^\alpha g_{\beta\overline{\sigma}} + \delta_\beta{}^\alpha R_{\rho\overline{\sigma}} + \delta_\rho{}^\alpha R_{\beta\overline{\sigma}}) \\ &+ \frac{R}{(n+1)(n+2)}(\delta_\beta{}^\alpha g_{\rho\overline{\sigma}} + \delta^\alpha_\rho g_{\beta\overline{\sigma}}) \end{aligned}$$

with $R_{\rho\overline{\sigma}} = R_\alpha{}^\alpha{}_{\rho\overline{\sigma}}$, $R = R_\rho{}^\rho$.

For dimensions $2n+1 \geq 5$ the strictly pseudoconvex integrable CR manifold M is then locally equivalent to the Heisenberg group if and only if the pseudoconformal curvature tensor S vanishes.

For dimension 3, Cartan's result gives a fifth order relative CR invariant. This means that the complex dilatation μ has to satisfy a certain fifth order differential equation in order for N_μ to be locally equivalent to N. This differential equation can be calculated explicitly.

3 The Fourier Transform Method

For the lowest dimensional case, the embedding problem seems to be particularly difficult. Jacobowitz and Trèves [8] show that the complex dilatations μ such that the CR manifold N_μ is not embeddable are dense. More precisely, assume that f^1 and f^2 are C^1-solutions of the Beltrami equation $\overline{Z}f = \mu Z f$ satisfying $df^1 \wedge df^2 \neq 0$. Then locally $f = (f^1, f^2)\colon N \to \mathbf{C}^2$ defines an embedding of the CR manifold N_μ into $\mathbf{C}^2$. In this situation, there exists a real vector field Q and a C^∞-function g vanishing to infinite order at a prescribed point $p \in N$ such that any two solutions f^1 and f^2 of the perturbed equation $\overline{Z}f = \mu Z f + gQf$ necessarily satisfy $df^1 \wedge df^2 = 0$ at $p \in N$.

In the example of Nirenberg [13] the perturbed Lewy operator does not admit any homogeneous solutions at all, except constants.

The question therefore arises, for which μ the Beltrami equation $\overline{Z}f = \mu Z f$ has nonconstant solutions.

In this section we will exhibit a class of functions μ with the property that the Beltrami equation $\overline{Z}f = \mu Z f$ on the lowest dimensional Heisenberg group $N = N^1$ admits nonconstant solutions. It will be assumed throughout that μ is a measurable function with compact support satisfying

$$\|\mu\|_\infty = \operatorname*{ess\,sup}_{p \in N} |\mu(p)| = \kappa < 1.$$

As in the classical case, the existence proof is based on the integral representation for solutions of the homogeneous equation $\overline{Z}f = g$. For g in the Schwartz class $\mathcal{S}(N)$, this equation only has a solution if the integrability condition

$$\overline{S}g(u) = g * \overline{S}(u) \quad \Big(= \int_N \overline{S}(v^{-1}u)g(v)\,dv\Big) = 0$$

with

$$\overline{S}(z,t) = \frac{1}{\pi^2(|z|^2 + it)^2} = \frac{-1}{\pi^2(t - i|z|^2)^2}$$

is satisfied (g has to be orthogonal to the kernel of Z, i.e. to the space of boundary values of antiholomorphic functions).

The canonical solution (i.e. the one orthogonal to the kernel of $\overline{Z}$) can then be represented in the form $f = g * k$ with

$$k(z,t) = -\frac{1}{\pi^2}\frac{\bar{z}}{(t + i|z|^2)(t - i|z|^2)}$$

(see Greiner-Kohn-Stein [7]). Furthermore, differentiation in the direction of Z shows that for $f \in \mathcal{S}(N)$, $Zf = \overline{Z}f * b$ with

$$b(z,t) = Zk(z,t) = \frac{2i}{\pi^2}\frac{\bar{z}^2}{(t + i|z|^2)(t - i|z|^2)}$$

This kernel is homogeneous of degree -4 and satisfies the cancellation property

$$\int_{\|(z,t)\|=1} b(z,t)\, d\sigma = 0$$

The associated convolution operator $B: \rightarrow g * b$ therefore extends to a bounded operator on all L^p-spaces, $1 < p < \infty$ (Korányi-Vági [11]). Similarly, the convolution operator $K: g \rightarrow g * k$ is homogeneous of degree $\lambda = -3$. It extends to a bounded mapping $K: L^p(N) \rightarrow L^q(N)$ if $1/q = 1/p - 1/4$ and $1 < p < q < \infty$ (Folland-Stein [5, p. 448]).

If now f is a solution of the Beltrami equation $\overline{Z}f = \mu Z f$ then outside suppμ, f is the boundary value of a holomorphic function ($\overline{Z}f = 0$). Therefore, if nontrivial solutions f of the Beltrami equation have to be found, it is natural to impose an asymptotic condition on f by prescribing holomorphic behavior at ∞.

This can be achieved by specifying a holomorphic function h (i.e. a smooth function satisfying $\overline{Z}h = 0$) and requiring $f - h$ to lie in the Sobolev space $W_1^p(N)$ of locally integrable functions

$$W_1^p(N) = \{g \in L^1_{\text{loc}}(N) = Zg, \overline{Z}g \in L^p\}.$$

The typical examples in connection with quasiconformal mappings are $h(z,t) = z$ and $h(z,t) = t + i|z|^2$.

Set now $g = \overline{Z}f$ and insert the expression

$$Zf = Z(f-h) + Zh = Bg + Zh$$

in the Beltrami equation so as to obtain

$$g = \mu B g + \mu Z h.$$

This equation has the formal solution

$$g = \sum_0^\infty (\mu B)^n (\mu Z h).$$

Since it is assumed that $\mu \in L^\infty(N)$ and that supp$\,\mu$ is compact, the function $\mu Z h$ is in $L^p(N)$. The sum then converges in $L^p(N)$ if the operator norm of μB (as an operator in $L^p(N)$) is smaller than 1. A solution to the Beltrami equation can then be obtained if g satisfies the integrability condition $\overline{S}g = 0$. In this case, the solution is given by $f = Kg + h$ and $f - h = W_1^p(N)$.

In general, the integrability condition $\overline{S}g = 0$ will not hold. This accounts for the nonembeddable CR manifold N_μ. However, by using the Fourier transform, it is possible to single out a class of functions μ for which the present method will provide nontrivial solutions for the Beltrami equation.

Observe finally that it is not necessary to require that μ have compact support. It suffices to have $\mu Z h \in L^p$.

As in the Euclidean case, the convolution operators S, K and B can be represented as multipliers on the Fourier transforms. It is our aim to calculate these multipliers. Explicit expressions for the matrix elements occurring in the Fourier transforms will then be used to control the operator μB occurring in the solution of the Beltrami equation. This will be possible at least for certain μ.

The Fourier transform on the Heisenberg group is well known. Our formulas are based on Faraut [4].

The Bargmann representation $(\mathfrak{H}_\lambda, T_\lambda)$ is the unitary representation on the space $\mathfrak{H}_\lambda$ of holomorphic function in $\mathbb{C}$ with scalar product

$$(\phi, \psi) = \frac{|\lambda|}{\pi} \int_{\mathbb{C}} \exp(-|\lambda|\,|\zeta|^2)\phi(\zeta)\overline{\psi}(\zeta)\,d\zeta$$

which is given by

$$T_\lambda(z,t)\phi(\zeta) = \begin{cases} \exp(-\lambda/2(it + |z|^2) - \lambda\bar{z}\zeta)\phi(\zeta + z) & \lambda > 0, \\ \exp(-\lambda/2(it - |z|^2) + \lambda z\zeta)\phi(\zeta + \bar{z}) & \lambda < 0. \end{cases}$$

Note that $T_\lambda(z,t) = T_{-\lambda}(\bar{z}, -t)$.

The induced representation of the Lie algebra is

$$\begin{array}{llll} T_\lambda(Z) = \frac{d}{d\zeta}, & T_\lambda(\overline{Z}) = -\lambda\zeta, & T_\lambda(T) = -\frac{i\lambda}{z} & \lambda > 0 \\ T_\lambda(Z) = \lambda\zeta, & T_\lambda(\overline{Z}) = \frac{d}{d\zeta}, & T_\lambda(T) = -\frac{i\lambda}{z} & \lambda < 0 \end{array}$$

By direct calculation it can be verified that the representations are unitary.

Observe that the present formulas slightly differ from the formulas in [4], since a different multiplication law for the Heisenberg group is used.

For a function f in the Schwartz class $\mathcal{S}(H)$ on the Heisenberg group, the Fourier transform at λ $(0 \neq \lambda \in \mathbb{R})$ is the operator on $\mathfrak{H}_\lambda$ given by

$$T_\lambda(f)\phi = \int_H f(z,t)T_\lambda(z,t)\phi\,dx\,dy\,dt$$

this operator is of trace class, its norm (the "trace norm") is defined by

$$||T_\lambda(f)||^2 = \operatorname{tr} T_\lambda^*(f)T_\lambda(f)$$

The inversion formula and the Plancherel formula for the Fourier transform are ([4])

$$f(z,t) = \frac{1}{4\pi^2} \int_{\mathbb{R}} \operatorname{tr} T_\lambda^*(z,t)T_\lambda(f)|\lambda|\,d\lambda$$

and

$$||f||^2 = \frac{1}{4\pi^2} \int_{\mathbb{R}} ||T_\lambda(f)||^2 |\lambda|\,d\lambda$$

where $||f||$ is the L^2-norm of f. On the basis of these formulas the Fourier transform can be extended to functions in $L^2(N)$.

Similarly to the Euclidean situation, the Fourier transform of the convolution

$$f * g(w) = \int_N f(v)g(v^{-1}w)\,dv \qquad f, g \in \mathcal{S}(N)$$

is given by the product

$$T_\lambda(f * g) = T_\lambda(f)T_\lambda(g).$$

We next calculate the matrix elements of the operators T_λ with respect to the orthonormal basis $\{e_n\} = \{\sqrt{|\lambda|^n/n!}\zeta^n\}$ in $\mathfrak{H}_\lambda$. For $\lambda > 0$ they are given by

$$\begin{aligned} t^\lambda_{n,m} &= (T_\lambda(z,t)e_m, e_n) \\ &= \frac{\lambda}{\pi}\sqrt{\frac{\lambda^n\lambda^m}{n!m!}} \int_{\mathbf{C}} \exp(-|\lambda|\,|\zeta|^2)\times \\ &\quad \times \exp(-\lambda/2(it + |z|^2) - \lambda\bar{z}\zeta)(\zeta + z)^m\bar{\zeta}^{-n}\,d\zeta \\ &= \frac{\lambda}{\pi}\sqrt{\frac{\lambda^n\lambda^m}{n!m!}} \exp(-i\lambda t/2)\exp(-\lambda|z|^2/2)\times \\ &\quad \times \int_{\mathbf{C}} \exp(-\lambda|\zeta|^2)\exp(-\lambda\bar{z}\zeta)(\zeta + z)^m\bar{\zeta}^n\,d\zeta. \end{aligned}$$

Denote the coefficient of ζ^n in the power series development of

$$\exp(-\lambda\bar{z}\zeta)(\zeta + z)^m$$

by c_{nm}. It can be determined from the expression

$$\exp(-\lambda\bar{z}\zeta)(\zeta + z)^m = \sum_{j=0}^{\infty} \frac{(-\lambda\bar{z}\zeta)^j}{j!} \sum_{k=0}^{m} \binom{m}{k} \zeta^k z^{m-k}$$

if the summation is only extended over the indices (j, k) such that $j + k = n$,

$$c_{nm}\zeta^n = \sum_{k=0}^{\min(n,m)} \frac{(-\lambda\bar{z}\zeta)^{n-k}}{(n-k)!} \binom{m}{k} \zeta^k z^{m-k}.$$

So if $m \geq n$ then

$$c_{nm} = \sum_{j=0}^{n} \frac{(-\lambda\bar{z})^j}{j!} \frac{m!}{(n-j)!(m-n+j)!} z^{m-n+j} = z^{m-n} L_n^{m-n}(\lambda|z|^2).$$

In this formula, the expression

$$L_n^\alpha(x) = \sum_{k=0}^{n} \frac{(-1)^k}{k!(n-k)!} \frac{\Gamma(n+\alpha+1)}{\Gamma(k+\alpha+1)} x^k \qquad \alpha > -1,\ x \geq 0$$

for the Laguerre polynomials is used. It can directly be derived from the definition

$$L_n^\alpha = \frac{1}{n!} e^x x^{-\alpha} \left(\frac{d}{dx}\right)^n (e^{-x} x^{n+\alpha}).$$

For $m \geq n$, the matrix elements are now

$$\begin{aligned} t^\lambda_{n,m}(z,t) &= \exp(-i\lambda t/2)\exp(-\lambda|z|^2/2)z^{m-n}L_n^{m-n}(\lambda|z|^2)\times \\ &\quad\times\sqrt{\frac{\lambda^n\lambda^m}{n!\,m!}}\frac{\lambda}{\pi}\int_{\mathbb{C}}\exp(-\lambda|\zeta|^2)\zeta^n\overline{\zeta^n}\,d\zeta \\ &= \sqrt{\frac{n!}{m!}}(\sqrt{\lambda}z)^{m-n}\exp(-i\lambda t/2)\exp(-\lambda|z|^2/2)L_n^{m-n}(\lambda|z|^2). \end{aligned}$$

whereas for $m < n$,

$$\begin{aligned} t^\lambda_{n,m}(z,t) &= (T_\lambda(z,t)e_m,e_n) = (e_m, T^*_\lambda(z,t)e_n) \\ &= \overline{(e_m, T_\lambda(-z,-t)e_n)} = \overline{t_{m,n}(-z,-t)} \\ &= \sqrt{\tfrac{m!}{n!}}(-\sqrt{\lambda}\,\bar{z})^{n-m}\times \\ &\quad\times\exp(-i\lambda t/2)\exp(-\lambda|z|^2/2)L_n^{n-m}(\lambda|z|^2). \end{aligned}$$

Finally, for $\lambda < 0$,

$$t^\lambda_{n,m}(z,t) = (T_\lambda(z,t)e_m,e_n) = (T_{-\lambda}(\bar{z},-t)e_m,e_n) = \overline{t^{|\lambda|}_{n,m}(z,t)}.$$

The differential operators Z, $\overline{Z}$ and T act as multiplication operators on the Fourier transform. In the case $\lambda > 0$ the relations

$$\begin{aligned} T_\lambda(Z)e_m &= \tfrac{d}{d\zeta}\sqrt{\tfrac{\lambda^m}{m!}}\zeta^m = \sqrt{\lambda m}e_{m-1} \\ T_\lambda(\overline{Z})e_m &= -\lambda\sqrt{\tfrac{\lambda^m}{m!}}\zeta^{m+1} = -\sqrt{\lambda(m+1)}e_{m+1} \end{aligned}$$

give

$$\begin{aligned} Zt^\lambda_{n,m} &= (T_\lambda(z,t)T_\lambda(Z)e_m,e_n) = \sqrt{\lambda m}t^\lambda_{n,m-1} \\ \overline{Z}t^\lambda_{n,m} &= -\sqrt{\lambda(m+1)}t^\lambda_{n,m+1} \\ Tt^\lambda_{n,m} &= -\tfrac{i\lambda}{2}t^\lambda_{n,m}. \end{aligned}$$

The corresponding formulas for $\lambda < 0$ are

$$\begin{aligned} Zt^\lambda_{n,m} &= \sqrt{|\lambda|(m+1)}t^\lambda_{n,m+1} \\ \overline{Z}t^\lambda_{n,m} &= \sqrt{|\lambda|m}\,t^\lambda_{n,m-1} \\ Tt^\lambda_{n,m} &= -\tfrac{i\lambda}{2}t^\lambda_{n,m}. \end{aligned}$$

The action of Z and $\overline{Z}$ on the Fourier transform can now be described by a matrix multiplication. Set

$$M(\lambda) = \sqrt{|\lambda|}\begin{pmatrix} 0 & \sqrt{1} & & & \\ & 0 & \sqrt{2} & & \\ & & 0 & \sqrt{3} & \\ & & & 0 & \ddots \\ & & & & \ddots \end{pmatrix} = \sqrt{|\lambda|}(\sqrt{k}\delta_{k,j+1})$$

and we use $M^*(\lambda)$ to denote the transposed matrix. If the operator $T_\lambda(z,t)$ is represented as a matrix $(t^\lambda_{n,m}(z,t))$ then the above equations show that

$$\left.\begin{array}{rcl} ZT_\lambda(z,t) &=& T_\lambda(z,t)\cdot M^*(\lambda) \\ \overline{Z}T_\lambda(z,t) &=& -T_\lambda(z,t)\cdot M(\lambda) \end{array}\right. \quad \text{if } \lambda > 0,$$

and

$$\left.\begin{array}{rcl} ZT_\lambda(z,t) &=& T_\lambda(z,t)\cdot M(\lambda) \\ \overline{Z}T_\lambda(z,t) &=& T_\lambda(z,t)\cdot M^*(\lambda) \end{array}\right. \quad \text{if } \lambda < 0.$$

These are the formulas which were derived by Greiner [6].

From now on, the Fourier transform $T_\lambda(f)$ of a function f at λ will also be represented as a matrix $\hat{f}(\lambda)$ with coefficients $\hat{f}^\lambda_{m,n}$. The inversion formula leads to the integral representations for Zf and $\overline{Z}f$:

$$\begin{array}{rcl} Zf(z,t) &=& \frac{1}{4\pi^2}\int\limits_0^\infty \operatorname{tr} T^*_\lambda(z,t)\hat{f}(\lambda)M(\lambda)|\lambda|\,d\lambda \\ && +\frac{1}{4\pi^2}\int\limits_{-\infty}^0 \operatorname{tr} T^*_\lambda(z,t)\hat{f}(\lambda)M^*(\lambda)|\lambda|\,d\lambda \\ \overline{Z}f(z,t) &=& -\frac{1}{4\pi^2}\int\limits_0^\infty \operatorname{tr} T^*_\lambda(z,t)\hat{f}(\lambda)M^*(\lambda)|\lambda|\,d\lambda \\ && +\frac{1}{4\pi^2}\int\limits_{-\infty}^0 \operatorname{tr} T^*_\lambda(z,t)\hat{f}(\lambda)M(\lambda)|\lambda|\,d\lambda \end{array}$$

The multipliers $\widehat{\widehat{S}}$, $\hat{K}$ and $\hat{B}$ associated to the integral operators can be read off these formulas (cf. Greiner [6], where this calculus was developed). Thus $\widehat{\widehat{S}}$ is given by

$$\widehat{\widehat{S}}(\lambda) = 0 \qquad\qquad \lambda > 0,$$

$$\widehat{\widehat{S}}(\lambda) = \begin{pmatrix} 1 & & & \\ & 0 & & \\ & & 0 & \\ & & & \cdots \end{pmatrix} := C \qquad \lambda < 0.$$

The matrix

$$M^{-1}(\lambda) = \frac{1}{\sqrt{|\lambda|}} \begin{pmatrix} 0 & & & & \\ \sqrt{1} & 0 & & & \\ & \sqrt{\frac{1}{2}} & 0 & & \\ & & \sqrt{\frac{1}{3}} & 0 & \\ & & & \ddots & \ddots \end{pmatrix}, \qquad \lambda \neq 0$$

is a right inverse for M, but only a partial left inverse. The relations

$$\begin{array}{rcl} MM^{-1} &=& I \\ M^{-1}M &=& I - C \\ M^*(M^{-1})^* &=& I - C \\ (M^{-1})^*M^* &=& I \end{array}$$

hold. If $g \in \mathcal{S}(N)$ satisfies the integrability condition $\hat{g}\widehat{\overline{S}} = 0$ then the function

$$f(z,t) = -\frac{1}{4\pi^2}\int_0^\infty \operatorname{tr} T^*_\lambda(z,t)\hat{g}(\lambda)(M^{-1})^*(\lambda)|\lambda|\,d\lambda + \frac{1}{4\pi^2}\int_{-\infty}^0 \operatorname{tr} T^*_\lambda(z,t)\hat{g}(\lambda)M^{-1}(\lambda)|\lambda|\,d\lambda$$

is the canonical solution (i.e. orthogonal to the kernel of $\overline{Z}$) of the inhomogeneous equation $\overline{Z}f = g$. The Fourier transform of the integral operator K is thus

$$\hat{K}(\lambda) = \begin{cases} -(M^{-1})^*\lambda & \lambda > 0, \\ M^{-1}(\lambda) & \lambda < 0. \end{cases}$$

Finally, the Fourier transform of B is

$$\hat{B}(\lambda) = \begin{cases} -(M^{-1})^* M & \lambda > 0, \\ M^{-1}M^* & \lambda < 0. \end{cases}$$

The coefficients are

$$\hat{b}^\lambda_{m,n} = \begin{cases} -\delta_{m+2,n}\sqrt{\frac{m+1}{m}} & \lambda > 0, \\ \delta_{m-2,n}\sqrt{\frac{m-2}{m-1}} & \lambda < 0. \end{cases}$$

In particular, the operator norm of $B: L^2(N) \to L^2(N)$ is $\sqrt{2}$. If however B is restricted to the subspace

$$L^2_-(N) = \{f \in L^2(N) : \hat{f}(\lambda) = 0 \text{ for a.e. } \lambda > 0\}$$

then the norm of $B: L^2_-(N) \to L^2_-(N)$ is 1. Clearly, both $L^2_-(N)$ and its orthogonal complement $L^2_+(N)$ are invariant under B.

Under the rotation group the space $L^2(N)$ decomposes into the mutually orthogonal subspaces

$$U^k = \{f \in L^2(N) : f(ze^{i\phi}, t) = e^{ik\phi}f(z,t)\}, \qquad k \in \mathbf{Z}$$

(with $z = |z|e^{i\phi}$). From the explicit form of the matrix elements of the representations it follows that the Fourier coefficients $\hat{f}^\lambda_{m,n}$ of the element $f \in U^k$ satisfy $\hat{f}^\lambda_{m,n} = 0$ for a.a. $\lambda > 0$ if $m - n \neq k$, $\hat{f}^\lambda_{m,n} = 0$ for a.a. $\lambda < 0$ if $n - m \neq k$, and B maps U^k into U^{k+2}. The complete orthogonal sums

$$D_j = \widehat{\bigoplus_{k \le j}} U^k$$

then obey the following multiplication law: $f \cdot \mu \in D_{j+m}$ if $f \in D_j$ and $\mu \in D_m \cap L^\infty(N)$. Furthermore the spaces $L^2_+(N)$ and $L^2_-(N)$ are multiplication invariant in the sense that $f \cdot \mu \in L^2_+(N)$ if $f \in L^2_+(N)$ and $\mu \in L^2_+(N) \cap L^\infty(N)$ (and similarly for $L^2_-(N)$).

THEOREM *Assume that h is holomorphic ($\overline{Z}h = 0$) and $\mu \in L^\infty(N)$. If either of the conditions*

1) $\mu \in L^2_+(N)$, $\mu Zh \in L^2_+(N)$ and $\|\mu\|_\infty < \frac{1}{\sqrt{2}}$,

2) $\mu \in D_{-2}$, $\mu Zh \in D_{-1}$ and $\|\mu\|_\infty < \frac{1}{\sqrt{2}}$,

3) $\mu \in D_{-2} \cap L^2_-(N)$, $\mu Zh \in D_{-1} \cap L^2_-(N)$ *and* $\|\mu\|_\infty < 1$,
is satisfied then the Beltrami equation $\overline{Z} f = \mu Z f$ *has a unique solution* f *such that* $f - h = W_1^2(N)$.

For the proof we have to show that the formal series

$$g = \sum_0^\infty (\mu B)^n (\mu Zh)$$

converges in $L^2(N)$ and satisfies the integrability condition $\overline{S} g = 0$.

In the first two cases, the condition $\|\mu\|_\infty < 1/\sqrt{2}$ implies the convergence of the series in $L^2(N)$. The norm of $B: L^2(N) \to L^2(N)$ is $\sqrt{2}$ and the operator norm of μB is thus smaller than 1.

In the third case, the operator B must be restricted to the invariant subspace $L^2_-(N)$. The condition $\mu \in D_{-2} \cap L^2_-(N)$ then implies that $L^2_-(N)$ is also invariant under multiplication by μ. When restricted to $L^2_-(N)$ the norm of μB is smaller than 1.

If $\mu \in D_{-2} \cap L^2_-(N)$ then the spaces D_k are invariant under the operator μB, since B maps D_k into D_{k+2} and the multiplication operator maps D_{k+2} into D_k. If in addition $\mu \in L^2_-(N)$ then $D_{-2} \cap L^2_-(N)$ is invariant under μB.

This shows that, in all three cases of the theorem, the function $g = \sum_0^\infty (\mu B)^n (\mu Zh)$ lies in the same space as μZh. But this space is chosen in such a way that all its elements satisfy the integrability condition, and so the proof is finished.

We finally remark that B is bounded on all L^p-spaces ($1 < p < \infty$). Therefore the series $\sum_0^\infty (\mu B)^n$ converges for some $p > 4$ if $\|\mu\|_\infty$ is small enough. In this case the solutions f of the Beltrami equation will be locally Hölder continuous (as a consequence of the Sobolev inequalities).

Acknowledgements: The authors are indebted to S. Webster for his patience in explaining some of the concepts used in the second section.

A. Korányi is partially supported by the National Science Foundation (DMS 8701530) and H. M. Reimann is partially supported by the Swiss National Foundation.

4 References

[1] T. Akahori. A new approach to the local embedding theorem of CR structures for $n \geq 4$. *Memoirs of the Amer. Math. Soc.* **366** (1987).

[2] E. Cartan. Sur la géometrie pseudo-conforme des hypersurfaces de l'espace de deux variables complexes. *Annali di Matematica* **11** (1932), 17–90. (Oeuvres II, 1231–1304).

[3] S. S. Chern & J. Moser. Real hypersurfaces in complex manifolds. *Acta Math.* **133** (1974), 219–271.

[4] J. Faraut. Analyse harmonique et fonctions spéciales. In *Deux cours d'analyse harmonique: École d'été d'analyse harmonique de Tunis, 1984*. Birkhäuser, 1987.

[5] G. B. Folland & E. M. Stein. Estimates for the $\bar{\partial}_b$ complex and analysis on the Heisenberg group. *Comm. Pure Appl. Math.* **27** (1974), 429–522.

[6] P. C. Greiner. On the Laguerre calculus of left-invariant convolution operators on the Heisenberg group. *Exposé XI, Séminaire Goulaouic-Meyer-Schwartz* (1980–81).

[7] P. C. Greiner, J. J. Kohn, & E. M. Stein. Necessary and sufficient conditions for the solvability of the Lewy equation. *Proc. Nat. Acad. of Sciences, USA* **72** (1975), 3287–3289.

[8] H. Jacobowitz & F. Trèves. Nonrealizable CR structures. *Inventiones Math.* **66** (1982), 231–249.

[9] J. J. Kohn. Boundaries of complex manifolds. In *Proc. Conf. Complex Analysis 1964, Mineapolis.* Springer-Verlag, 1965.

[10] A. Korányi & H. M. Reimann. Quasiconformal mappings on the Heisenberg group. *Inventiones Math.* **80** (1985), 309–338.

[11] A. Korányi & S. Vági. Singular integrals on homogeneous spaces and some problems of classical analysis. *Annali Norm. Sup. Pisa* **25** (1971), 575–648.

[12] M. Kuranishi. Strongly pseudoconvex CR structures on small balls I, II, III. *Ann. Math.* **115** (1982), 451–500; **116** (1982), 1–64; **116** (1982), 249–330, 1982.

[13] L. Nirenberg. *Lectures on linear partial differential equations*, CMBS **17**. Providence, RI: Amer. Math. Soc., 1973.

[14] S. Webster. Pseudo-Hermitian structures on a real hypersurface. *J. Diff. Geom.* **13** (1978), 25–41.

[15] ——. Analytic discs and the regularity of CR mappings of real submanifolds. *Proc. Symp. in Pure Math.* **41** (1984), 199–208.

[16] ——. On the proof of Kuranishi's embedding theorem. Preprint.

On the stability of positive semigroups generated by operator matrices

by

RAINER NAGEL

Mathematisches Institut der Universität Tübingen,
Auf der Morgenstelle 10, D–7400 Tübingen

Dedicated to E. Vesentini on the occasion of his 60th *birthday.*

Abstract. We consider unbounded operator matrices generating positive semigroups on products of Banach lattices. Generalizing the concept of an *M–matrix* (see [2]) we characterize the stability of the generated semigroup by simple criteria.

1. Introduction

In many, but not all cases (see [12], A–III and A–IV) a strongly continuous semigroup $(T(t))_{t\geq 0}$ of bounded linear operators on a Banach space is *uniformly exponentially stable*, i.e. $\lim_{t\to\infty} e^{-\epsilon t}\|T(t)\| = 0$ for some $\epsilon > 0$, if and only if the *spectral bound*

$$s(A) := \sup\{\mathrm{Re}\lambda : \ \lambda \in \sigma(A)\}$$

of its generator A satisfies

$$s(A) < 0.$$

For example, this holds for positive semigroups on the Banach lattices $C_0(X)$, $L^1(\mu)$ and $L^2(\mu)$ ([12], B–IV, Thm. 1.4, C–IV, Thm. 1.1 and [8]).

In addition, the Perron–Frobenius spectral theory for positive semigroups facilitates considerably the determination of $s(A)$. As one of the fundamental spectral properties of positive semigroups we mention that the spectral bound $s(A)$ always belongs to the spectrum $\sigma(A)$ ([12], C–III, Thm. 1.1), hence one obtains

$$s(A) = \sup\{\lambda : \ \lambda \in \sigma(A) \cap \mathbb{R}\}.$$

For positive semigroups arising from delay equations [7] or population equations [9] these facts yielded simple and useful stability criteria.

On the other hand, for positive matrix semigroups on the Banach lattice $\mathbb{R}^n$ one has the following well known characterization of stability (see [2] or [10]).

Theorem. Let $A = (a_{ij})_{n\times n}$ be a real matrix satisfying $a_{ij} \geq 0$ for $i \neq j$. Then the semigroup $(e^{At})_{t\geq 0}$ generated by A is positive and the following assertions are equivalent.

(a) $(e^{At})_{t\geq 0}$ is uniformly exponentially stable.

(b) $s(A) < 0$.

(c) $\det A \neq 0$ and $A^{-1} \leq 0$.

(d) All principal minors of $-A$ are greater than zero.

The purpose of this paper is to extend the above theorem to systems of evolution equations, i.e. to semigroups on products of infinite dimensional Banach spaces. At the same time our proof immediately suggests an algorithm for possible numerical applications. We use the basic results from the theory of one–parameter semigroups (see [1], [6] or [14]) and refer to [4], [12] or [15] for additional information on positive operators and positive semigroups.

2. Systems of evolution equations (known results)

Consider a system of linear evolution equations of the form

$$(SE) \qquad \dot{v}_i(t) = \sum_{j=1}^{n} A_{ij} v_j(t), \quad v_i(0) = f_i \text{ for } i = 1, \dots, n$$

for functions $v_i(\cdot)$ on $\mathbb{R}_+$ with values in (possibly different) Banach spaces E_i and for linear (possibly unbounded) operators A_{ij} from E_j into E_i. Using the conventions from linear algebra this system can be written as

$$(SE) \qquad \dot{u}(t) = \mathcal{A}u(t), \quad u(0) = u_0$$

for a function $u(\cdot)$ with values in $\mathcal{E} := E_1 \times \cdots \times E_n$ and an operator matrix $\mathcal{A} := (A_{ij})_{n \times n}$. In the following we always assume that the operators A_{ii} with domains $D(A_{ii})$ are generators of strongly continuous semigroups $(T_i(t))_{t\geq 0}$ on E_i and that A_{ij} are bounded for all $i \neq j$. Then the operator matrix $\mathcal{A}$ with domain $D(\mathcal{A}) := D(A_{11}) \times \cdots \times D(A_{nn})$ generates a strongly continuous semigroup $(\mathcal{T}(t))_{t\geq 0}$ on $\mathcal{E}$ (use [6], Thm. I. 6. 4), (SE) is well posed in the sense of [6] and the solutions are given by the semigroup. For results on operator matrices with unbounded off–diagonal entries we refer to [11].

We are now interested in exponential stability of the solutions of (SE), i.e. of the semigroup $(\mathcal{T}(t))_{t\geq 0}$. As explained above it suffices in many cases to determine the spectral bound $s(\mathcal{A})$ of the operator matrix $\mathcal{A}$. To that purpose we use the standard spectral theoretic notations (see [12], A–III) and the following results.

2.1 Characteristic operator functions (see [13]). Take $\mathcal{E} := E \times F$ to be the product of two Banach spaces E and F. Let $\mathcal{A} := \begin{pmatrix} A & B \\ C & D \end{pmatrix}$ be a 2×2 operator matrix on $\mathcal{E}$ with domain $D(\mathcal{A}) = D(A) \times D(D)$ and bounded off–diagonal elements B, C. For $\lambda \notin \sigma(A)$ and $R(\lambda, A) := (\lambda - A)^{-1}$ we define the *characteristic operator function*

$$\Delta(\lambda) := D + CR(\lambda, A)B$$

and have the following characterization of spectral values of $\mathcal{A}$ (not belonging to $\sigma(A)$):

$$\lambda \in \sigma(\mathcal{A}) \quad \Longleftrightarrow \quad \lambda \in \sigma(\Delta(\lambda)).$$

If $\lambda \notin \sigma(\mathcal{A}) \cup \sigma(A)$ then the resolvent $R(\lambda, \mathcal{A}) := (\lambda - \mathcal{A})^{-1}$ is given by the operator matrix

$$(*) \quad R(\lambda, \mathcal{A}) = \begin{pmatrix} R(\lambda, A + BR(\lambda, D)C) & R(\lambda, A + BR(\lambda, D)C)BR(\lambda, D) \\ R(\lambda, D + R(\lambda, A)B)CR(\lambda, A) & R(\lambda, D + CR(\lambda, A)B) \end{pmatrix}.$$

Next we are interested in systems (SE) having positive solutions for every positive initial value, or equivalently, in semigroups $(T(t))_{t\geq 0}$ leaving invariant a positive cone in $\mathcal{E}$. To that purpose we assume that the spaces E_i are Banach lattices (as, e.g., $C(X)$, $L^p(\mu)$) and consider the canonical product cone in $\mathcal{E}$. Then one of the many classical characterizations of positive matrix semigroups on $\mathbb{R}^n$ (see [5] or [2], [10]) can be generalized as follows.

2.2 Characterization of positive matrix semigroups (see [11]). For the operator matrix $\mathcal{A} = (A_{ij})_{n\times n}$ on the product $\mathcal{E} := E_1 \times \ldots \times E_n$ of Banach lattices E_i the following assertions are equivalent.

(a) The semigroup $(T(t))_{t\geq 0}$ generated by $\mathcal{A}$ is positive.

(b) (i) Each A_{ii}, $i = 1, \ldots, n$, generates a positive semigroup on E_i.

(ii) Each A_{ij}, $i \neq j$, is a positive operator from E_j into E_i.

3. Characterization of stability

We start with the study of 2×2–operator matrices and only then extend the results to matrices of arbitrary size.

Let E and F be two Banach lattices and take an operator matrix $\mathcal{A} = \begin{pmatrix} A & B \\ C & D \end{pmatrix}$ where A and D are generators while B and C are bounded. In addition, let A and D generate positive semigroups while B and C are positive. Then $\mathcal{A}$ generates a semigroup $(T(t))_{t\geq 0}$ on $\mathcal{E} := E \times F$ which is positive by 2.1.

In order to characterize the stability of the solutions of the system (SE) corresponding to $\mathcal{A}$ we show "$s(\mathcal{A}) < 0$". The subsequent considerations are based on results from the Perron–Frobenius theory of positive semigroups. In particular we need the following lemma (see [12], C–III, Thm.1.1).

3.1 Lemma. Let A be the generator of a strongly continuous semigroup of *positive* operators on some Banach lattice. Then

(i) $s(A) \in \sigma(A)$ and

(ii) $s(A) = \inf\{\lambda \in \mathbb{R} : R(\lambda, A) \geq 0\}$.

In particular, for positive semigroups one obtains immediately the following surprisingly simple stability criterium which is an abstract version of the maximum principle.

3.2 Stability Criterion. For the generator A of a positive semigroup on a Banach lattice the following assertions are equivalent.

(a) $s(A) < 0$,

(b) $0 \in \varrho(A)$ and $A^{-1} \leq 0$.

These results will now be applied to semigroups generated by the above operator matrix.

3.3 Theorem. For the operator matrix $\mathcal{A} = \begin{pmatrix} A & B \\ C & D \end{pmatrix}$ generating a positive semigroup on $E \times F$ the following assertions are equivalent.

(a) $s(\mathcal{A}) < 0$,

(a') $0 \in \varrho(\mathcal{A})$ and $\mathcal{A}^{-1} \leq 0$,

(b) $s(A) < 0$ and $s(D - CA^{-1}B) < 0$,

(b') $0 \in \varrho(A) \cap \varrho(D - CA^{-1}B)$ and $A^{-1} \leq 0$, $(D - CA^{-1}B)^{-1} \leq 0$,

(c) $s(D) < 0$ and $s(A - BD^{-1}C) < 0$,

(c') $0 \in \varrho(D) \cap \varrho(A - BD^{-1}C)$ and $D^{-1} \leq 0$, $(A - BD^{-1}C)^{-1} \leq 0$.

Proof. The equivalence of the primed and unprimed versions is an immediate consequence of the Stability Criterion 3.2. Since (c) is obtained from (b) by changing $E \times F$ to $F \times E$ (and vice versa) it suffices to show the equivalence of (a) and (b).

(a) $\Rightarrow$ (b). Since B und C are positive operators the semigroup generated by $\mathcal{A}$ dominates the semigroup generated by $\mathcal{A}_0 := \begin{pmatrix} A & 0 \\ 0 & D \end{pmatrix}$ (use the series expansion of the perturbed semigroup, e.g., [6], I. 6.5). The integral representation of the resolvent ([12], C–III, Thm. 1.2) yields $0 \leq R(\lambda, \mathcal{A}_0) \leq R(\lambda, \mathcal{A})$ for $\lambda > s(\mathcal{A}_0)$, hence $s(\mathcal{A}_0) \leq s(\mathcal{A})$. Therefore $s(\mathcal{A}) < 0$ implies $\sup(s(A), s(D)) = s(\mathcal{A}_0) < 0$ and $A^{-1} \leq 0$, $D^{-1} \leq 0$. From (a') we infer that $R(0, \mathcal{A}) = -\mathcal{A}^{-1}$ exists and is a positive operator on $E \times F$. Moreover we have from [13], Thm. 2. 4 (compare with (*) in 2. 1) the matrix representation

$$(**) \qquad -\mathcal{A}^{-1} = \begin{pmatrix} -A^{-1}(Id + B(D - CA^{-1}B)^{-1}CA^{-1} & A^{-1}B(D - CA^{-1}B)^{-1} \\ (D - CA^{-1}B)^{-1}CA^{-1} & -(D - CA^{-1}B)^{-1} \end{pmatrix}.$$

This implies that each entry and in particular $-(D - CA^{-1}B)^{-1}$ is a positive operator. Thus we have shown (b) and (b').

(b) $\Rightarrow$ (a). We use again the matrix representation (**) of the resolvent $R(\lambda, \mathcal{A})$ in $\lambda = 0$. In fact, (b) means that $-A^{-1}$ and $-(D - CA^{-1}B)^{-1}$ are positive. Therefore all entries in (**) exist and are positive which shows (a') and (a). ∎

In order to obtain a criterion for matrices $\mathcal{A} = (A_{ij})_{n \times n}$ on the product $\mathcal{E} := E_1 \times \cdots \times E_n$ of n Banach lattices we use the notation

$$\mathcal{A}_k := (A_{ij})_{k \times k}, \quad \mathcal{B}_k := \begin{pmatrix} A_{1k} \\ \vdots \\ A_{k-1,k} \end{pmatrix} \text{ and } \mathcal{C}_k := (A_{k1}, \dots, A_{k,k-1}).$$

As before we assume that the diagonal elements A_{ii} generate positive semigroups on E_i while the off-diagonal elements A_{ij}, $i \neq j$, are positive and bounded from E_j into E_i. Clearly, for $1 \leq k \leq n$ we can write

$$\mathcal{A}_1 := A_{11} \text{ and } \mathcal{A}_k = \begin{pmatrix} \mathcal{A}_{k-1} & \mathcal{B}_k \\ \mathcal{C}_k & A_{kk} \end{pmatrix}.$$

Here $\mathcal{A}_{k-1}$, resp., A_{kk} generate positive semigroups on $E_1 \times \cdots \times E_{k-1}$, resp., E_k and $\mathcal{B}_k$, resp., $\mathcal{C}_k$ are positive and bounded from E_k into $E_1 \times \cdots \times E_{k-1}$, resp., from $E_1 \times \cdots \times E_{k-1}$ into E_k. Therefore we can apply the above theorem to each of the 2×2-matrices $\mathcal{A}_k$, $1 < k \leq n$ and obtain the following corollary.

3.4 Corollary. For the above operator matrix $\mathcal{A} = (A_{ij})_{n \times n}$ generating a positive semigroup on $\mathcal{E} := E_1 \times \cdots \times E_n$ the following assertions are equivalent.

(a) $s(\mathcal{A}) < 0$,

(a′) $0 \in \varrho(\mathcal{A})$ and $\mathcal{A}^{-1} \leq 0$,

(b) $s(A_{11}) < 0$ and $s(A_{kk} - \mathcal{C}_k \mathcal{A}_{k-1}^{-1} \mathcal{B}_k) < 0$ for $k = 2, \ldots, n$,

(b′) $0 \in \varrho(A_{11}) \cap \varrho(A_{kk} - \mathcal{C}_k \mathcal{A}_{k-1}^{-1} \mathcal{B}_k)$ and $A_{11}^{-1} \leq 0$, $(A_{kk} - \mathcal{C}_k \mathcal{A}_{k-1}^{-1} \mathcal{B}_k)^{-1} \leq 0$ for $k = 2, \ldots, n$.

In order to evaluate the above stability criterion we state the following observations.

3.5 Remark. (i) The stability of the semigroup generated by $\mathcal{A}$ on the product space $\mathcal{E} = E_1 \times \cdots \times E_n$ can be determined by properties of certain operators on the (smaller) factor spaces E_k.
(ii) Given the inverse of $A_{kk} - \mathcal{C}_k \mathcal{A}_{k-1}^{-1} \mathcal{B}_k$ as in (b′) there are explicit formulas for $\mathcal{A}_k^{-1}$ (see (**)). Hence criterion (b), resp. (b′) is of algorithmic nature.

4. Applications

Our abstract criterion will now be applied to certain more concrete situations. First we retrieve some classical results on scalar "M–matrices" (see [5] or [2], [10]).

4.1 Scalar Matrices. The real matrix $\mathcal{A} = (a_{ij})_{n \times n}$ generates a positive semigroup on $\mathbb{R}^n$ if and only if $a_{ij} \geq 0$ for $i \neq j$. This semigroup is exponentially stable if and only if

(a) $s(\mathcal{A}) < 0$,

which by Cor. 3.4 is equivalent to

(b) $a_{11} < 0$ and $a_{kk} - \mathcal{C}_k \mathcal{A}_{k-1}^{-1} \mathcal{B}_k < 0$ for $k = 2, \ldots, n$.

Since $\det \mathcal{A}_k = (a_{kk} - \mathcal{C}_k \mathcal{A}_{k-1}^{-1} \mathcal{B}_k) \cdot \det \mathcal{A}_{k-1}$ we obtain the wellknown stability condition using the principal minors of $\mathcal{A}$:

(c) $(-1)^{k+1} \cdot \det \mathcal{A}_k < 0$ for $k = 1, \ldots, n$.

4.2 Reaction–Diffusion Systems. Let Ω be a region in $\mathbb{R}^d$ and consider the Laplacian Δ on $E := C_0(\Omega)$ with appropriate domain $D(\Delta)$ such that Δ generates a positive semigroup on E. Choose operators $A_{ij} \in \mathcal{L}(E)$, $1 \le i,j \le n$, satisfying $0 \le A_{ij}$ for $i \ne j$ and $0 \le e^{tA_{ii}}$ for all $t \ge 0$ (compare [12], C–II, Thm. 1. 11).

Many linear reaction–diffusion systems (e.g., [3]) can now be expressed by an operator matrix

$$\mathcal{A} := \operatorname{diag}(a_i \cdot \Delta)_{n\times n} + (A_{ij})_{n\times n}$$

on the product space $\mathcal{E} := E^n$ and with coefficients $a_i \in \mathbb{R}_+$. Clearly, $\mathcal{A}$ generates a positive semigroup on $\mathcal{E}$ which is uniformly exponentially stable if and only if $s(\mathcal{A}) < 0$ (use [12], B–IV, Thm. 1. 4). By Cor. 3.4 this is equivalent to the property that certain bounded perturbations of $a_k \cdot \Delta$, $k = 1, \dots, n$, have negative spectral bound.

In the simple case that $A_{ij} = b_{ij} \cdot Id$, b_{ij} real, this becomes equivalent to

$$s(\operatorname{diag}(a_i \cdot \sigma)_{n\times n} + (b_{ij})_{n\times n}) < 0, \qquad (*)$$

where $\sigma := s(\Delta)$. Hence we can apply the criterion from 4.1.

4.3 Semigroups on $L^2 \times \mathbb{R}$. In many situations (e.g. [16] or [13], Ex. 2. 5) one studies semigroups on product spaces where one factor space is one or finite dimensional.

In order to apply our stability criterion we assume here that $E = L^2(\mu)$ and that A with domain $D(A)$ generates a positive semigroup on $L^2(\mu)$. For positive functions $g, h \in L^2(\mu)$ and $\delta \in \mathbb{R}$ we consider

$$\mathcal{A} := \begin{pmatrix} A & g \\ h & \delta \end{pmatrix}, \quad \text{e.g.,} \quad \mathcal{A}\begin{pmatrix} f \\ x \end{pmatrix} := \begin{pmatrix} Af + \delta g \\ \int fh d\mu + \delta x \end{pmatrix}$$

with domain $D(\mathcal{A}) = D(A) \times \mathbb{R}$ which clearly generates a positive semigroup in $E \times \mathbb{R}$. Its stability is determined by the spectral bound $s(\mathcal{A})$ (use [8]) and hence $s(\mathcal{A}) < 0$ is equivalent to

$$s(A) < 0 \quad \text{and} \quad \delta - \int hA^{-1} g d\mu < 0.$$

REFERENCES

[1] A. BELLENI–MORANTE, "Applied Semigroups and Evolution Equations", Oxford University Press 1979.

[2] A. BERMAN AND J. PLEMMONS, "Nonnegative Matrices in the Mathematical Sciences", Academic Press, New York 1979.

[3] V. CAPASSO AND L. MADDALENA, Convergence to equilibrium states for a reaction–diffusion system modelling the spatial spread of a class of bacterial and viral diseases, J. Math. Biology **13** (1981), 173–184.

[4] PH. CLEMENT, H. J. A. M. HEIJMANS et al., "One–parameter Semigroups", CWJ Monographs **5**, North Holland, Amsterdam 1987.

[5] M. FIEDLER AND V. PTAK, On matrices with non–positive off–diagonal elements and positive principal minors, Czech. Math. J. **12** (1962), 382–400.

[6] J. A. GOLDSTEIN, "Semigroups of Linear Operators and Applications", Oxford University Press, New York 1985.

[7] A. GRABOSCH, Translation semigroups and their linearizations on spaces of integrable functions, Trans. Amer. Math. Soc. (to appear)

[8] G. GREINER AND R. NAGEL, On the stability of stongly continuous semigroups of positive operators on $L^2(\mu)$, Annali Scuola Normale Sup. Pisa **10** (1983), 257–262.

[9] G. GREINER AND R. NAGEL, Growth of cell populations via one–parameter semigroups of positive operators. In: J. A. Goldstein, S. Rosencrans, G. Sod (eds.): "Mathematics Applied to Science", Academic Press 1988, p. 79–105.

[10] H. MINC, "Nonnegative Matrices", Wiley–Interscience 1988.

[11] R. Nagel, Well–posedness and positivity for systems of linear evolution equations, Conferenze del Seminario di Matematica Bari **203** (1985), 1–29.

[12] R. NAGEL (ed.), "One–Parameter Semigroups of Positive Operators", Lecture Notes Math. **1184**, Springer–Verlag 1986.

[13] R. NAGEL, Towards a "matrix theory" for unbounded operator matrices, Math. Z. (to appear)

[14] A. PAZY, "Semigroups of Linear Operators and Applications to Partial Differential Equations", Springer–Verlag 1983.

[15] H. H. SCHAEFER, "Banach Lattices and Positive Operators", Springer–Verlag 1974.

[16] E. VESENTINI, Semigroups of holomorphic isometries, Adv. Math. **65** (1987), 272–306.

The Levi Problem on Algebraic Manifolds

by Raghavan Narasimhan

University of Chicago

For Edoardo Vesentini

In the early '60s, A. Andreotti and E. Vesentini began a systematic attempt to extend the vanishing theorems of Kodaira and Nakano to non-compact manifolds. This led, among other things, to their paper [1] on what is now often referred to as the $\overline{\partial}$-method. Their approach is more closely tied to differential geometry than the parallel, more analytic, method of Hörmander [5].

T. Napier and K. Pinney of the University of Chicago have used the $\overline{\partial}$-method to obtain new results concerning variants of two well-known problems in complex analysis. I shall attempt to state these results and to outline some very interesting open problems that arise from this work.

In what follows, if L is a line bundle and p an integer ≥ 1, L^p stands for the pth tensor power $L^{\otimes p}$ of L.

1. The Levi Problem.

It is well-known that (relatively compact) strongly pseudoconvex domains on any complex manifold are holomorph-convex. Howewer, weakly pseudoconvex domains, even with smooth (or analytic) boundary, need not carry any non-constant holomorphic functions (Grauert [3, 4]; see also [8]). The first result is an attempt to restore global convexity in this situation (at least on projective manifolds).

DEFINITION 1. Let $D \subset\subset M$ be a relatively compact open subset of a complex manifold M and let L be a holomorphic line bundle on M. We say that D is *convex with respect to* L if the following holds:

Let $a \in \partial D$, and U a neighbourhood of a in M on which there exists a holomorphic section $s \in \Gamma(U, L)$ without any zeros. Then, for any infinite sequence $\{x_\nu\}_{\nu \geq 1}$ of points $x_\nu \in D$, converging to a, there exists a holomorphic section $f \in \Gamma(D, L)$ such that the function f/s on $U \cap D$ is unbounded on the sequence $\{x_\nu\}_{\nu \geq \nu_0}$, ν_0 large.

One then has the following

Theorem 1.1: (K. Pinney) *Let D be a weakly pseudoconvex domain with smooth boundary on the projective manifold M, and let L be a positive (i.e., ample) line bundle on M. Then there exists an integer $p_0 > 0$ such that D is convex with respect to L^p for $p \geq p_0$.*

There are two main steps in the proof.

Step 1. Let D be a (relatively compact) weakly pseudoconvex domain in the Kähler manifold M. If D has smooth boundary, then there exists a complete Kähler metric on D.

The smoothness of ∂D can be replaced by the condition that ∂D be C^2 (both in Step 1 and Theorem 1.1). Theorem 1.1 is probably true without this assumption.

Step 1 enables one to use the $\overline{\partial}$-method of Andreotti and Vesentini. Their basic remark may be formulated as follows: when the base manifold carries a complete Kähler metric, the *a priori* estimates needed to solve the equation $\overline{\partial} u = f$ need only be checked for C^∞ forms with compact support. This reduces the problem to finding lower bounds for $\langle \Delta u, u \rangle$, Δ being the Laplace operator associated with a metric on a line bundle, and this latter expression can be estimated in terms of curvature forms (Weitzenböck formula).

For the second step, let k be an integer such that for any $a \in \partial D$ there exist $s_1, \ldots, s_n \in \Gamma(M, L^k)$ (where $n = \dim_{\mathbb{C}} M$) which vanish at a and such that the hypersurfaces $s_j = 0$ intersect transversally.

Step 2. Let $\varphi \colon \mathcal{O}_M^n \to L^k$ be the morphism $\varphi(\alpha_1, \ldots, \alpha_n) = \sum \alpha_j s_j$. Then there exists an integer $l_0 > 0$ such that for every integer $l \geq l_0$ the image of $\varphi \otimes 1 \colon \Gamma(D, L^l) \to \Gamma(D, L^{l+k})$ contains all sections of the form $f|_D$, where $f \in \Gamma(M, L^{k+l})$ and vanishes at the common zeroes of $s_1, \ldots, s_n$ *inside* D.

Theorem 1.1 is an easy consequence of this result.

It seems very likely that one can use Theorem 1.1 to prove holomorph-convexity in certain cases. I shall state the following as a conjecture; special cases have already been proved.

CONJECTURE. Let M be a projective manifold with $c_1(M) \geq 0$. Let D be a pseudoconvex domain with smooth boundary on M. Suppose that there exists a complete Kähler metric g on D such that

$$-a\, g \leq \mathrm{Ric}(g) \leq -b\, g,$$

where a and b are positive constants. Then D is Stein.

This would enable one to show that several of the examples constructed by Grauert [4, 8] do not carry a complete Kähler-Einstein metric although they are pseudoconvex and do not contain compact analytic sets of positive dimension. One should also mention, in this connection, the theorem of Elencwajg on the holomorph-convexity of pseudoconvex domains on (not necessarily compact) manifolds M of positive holomorphic sectional curvature [2].

2. The Shafarevich Conjecture.

This conjecture is the following: let M be a projective algebraic manifold. The universal covering $\widetilde{M}$ of M is then holomorph-convex.

It should be pointed out at once that a similar statement for *arbitrary* coverings of M is false.

EXAMPLE. Consider the vectors $e_1 = (1,0)$, $e_2 = (0,1)$ and $e_3 = (\alpha,\beta)$ in $\mathbf{C}^2$. Let Λ be the subgroup generated by e_1, e_2 and e_3. If 1, α and β are linearly independent over $\mathbf{Z}$, then there is no non-constant holomorphic function f on $\mathbf{C}^2$ with $f(z+\lambda) = f(z)$ for every $\lambda \in \Lambda$; so there are no non-constant holomorphic functions on $\widetilde{M} = \mathbf{C}^2/\Lambda$. Clearly, such manifolds $\widetilde{M}$ occur as coverings of projective two-dimensional complex tori.

DEFINITION 2. Let M be a compact complex manifold and $\pi\colon \widetilde{M} \to M$ a covering manifold. Let L be a holomorphic line bundle on M. We say that $\widetilde{M}$ is *convex with respect to* L if the following holds:

Let $U_1, \ldots, U_N$ be open sets on M such that $L|_{U_j}$ is trivial and $M = \bigcup U_j$. Let $V_j \subset\subset U_j$ be open sets on M such that $\bigcup V_j = M$. Let $s_j \in \Gamma(U_j, L)$ be a holomorphic section without zeros on U_j $(j = 1, \ldots, N)$. Then, given an infinite discrete sequence $x_\nu \in \widetilde{M}$ $(\nu \geq 1)$, there exists $f \in \Gamma(\widetilde{M}, \pi^* L)$ and an index j such that the function $x \mapsto f(x)/s_j(\pi(x))$ on $\pi^{-1}(U_j)$ is unbounded on the (infinite) set $\{x_\nu \mid \pi(x_\nu) \in U_j\}$.

This is equivalent to requiring the following: for any hermitian metric h on L and any sequence $\{x_\nu\}$ as above, there exists $f \in \Gamma(\widetilde{M}, \pi^* L)$ such that $h^*\big(f(x_\nu)\big)$ is unbounded, where h^* is the metric on $\pi^* L$ induced by h.

T. Napier has proved the following theorem; I had proved special cases of the theorem for canonically ample varieties (in which case one can obtain explicit bounds for p_0 in terms of the dimension).

Theorem 2.1: (T. Napier) *Let M be a projective manifold, let $\pi\colon \widetilde{M} \to M$ be any covering manifold, and let L be a positive line bundle on M. Then there exists an integer p_0 such that $\widetilde{M}$ is convex with respect to L^p for $p \geq p_0$.*

The proof is based again on the $\overline{\partial}$-method which is used to construct sections of L^p with given values on a suitable subsequence of $\{x_\nu\}$. Napier's result is actually considerably stronger than Theorem 2.1; it deals with so-called "manifolds with bounded geometry" and the sections constructed have exponential growth.

We turn next to the Shafarevich conjecture itself. The results obtained so far are all in the case when M is a smooth projective surface (i.e., $\dim_{\mathbf{C}} M = 2$).

Theorem 2.2: *If M is a smooth, projective, elliptic surface, then its universal covering is holomorph-convex.*

This theorem was also proved (independently) by Gurjar and Shastry of the Tata Institute of Fundamental Research (unpublished).

The main steps in the proof are as follows.

Let $M \xrightarrow{\varphi} C$ be a holomorphic map onto a smooth curve C such that the generic fibre is a smooth elliptic curve. Let $\pi\colon \widetilde{M} \to M$ be the universal covering.

The first step is to show that if $\widetilde{M}$ is not compact and $F_1, \ldots, F_k$ are the singular fibres, then, if $k > 0$, $\pi^{-1}(M \setminus \bigcup_{l=1}^{k} F_l)$ is Stein. The proof of this is analogous to the so-called Serre conjecture over Riemann surfaces.

The main step is to show that there exist neighbourhoods U_l of F_l (where l ranges from 1 to k) such that $\pi^{-1}(U_l)$ is holomorph-convex. This is done by a case by case analysis using Kodaira's classification of singular fibres and their neighbourhoods. This result can also be proved using the techniques of Napier mentioned below in connection with Theorem 2.3.

The final step is a *Heftungslemma* of the kind used in the Serre conjecture (about fibrations with Stein fibres over a Stein base) in the case when base and fibre have dimension 1.

The other results known are due to T. Napier and can be summarized as follows:

Theorem 2.3: *Let M be a smooth projective surface, and let $\pi: \widetilde{M} \to M$ be the universal covering. Assume that the first Betti number $b_1(M)$ is non-zero, and that $\widetilde{M}$ does not contain any connected, non-compact 1-complex dimensional subspace all of whose irreducible components are compact. Then $\widetilde{M}$ is holomorph-convex.*

Napier's proof starts by considering the canonical map of X into its Albanese variety $\mathrm{Alb}(X)$: $X \xrightarrow{\alpha} \mathrm{Alb}(X)$. If $b_1(X) > 0$, then α is non constant. Consider the singular fibres of α when $\dim \alpha(X) = 1$, and the fibres of positive dimension (which get mapped onto a point in $\mathrm{Alb}(X)$) when $\dim \alpha(X) = 2$. Let $C_1, \ldots, C_N$ be the connected components of these curves. One has only to assume that if $\pi: \widetilde{X} \to X$ is the universal covering of X, no connected component of $\pi^{-1}(C_i)$ is non-compact but has only compact irreducible components. It is not hard to show that $\pi^{-1}(X \setminus \bigcup C_i)$ is holomorph-convex (in fact Stein except in certain simple cases). The main difficulty is in the proof of the following result:

Let X be a smooth surface and let $p: \widehat{X} \to X$ be *any* covering manifold. Let C be a connected curve on X. Assume that if $p^{-1}(C)$ has a non-compact *connected* component $\widetilde{C}$, then $\widetilde{C}$ has at least one non-compact *irreducible* component. Then, there exists an open set $U \supset C$ such that $p^{-1}(U)$ is holomorph-convex.*

The proof of this result involves a subtle construction of plurisubharmonic functions. The essence is contained in the following result.

Let $p: \widehat{X} \to X$ be a covering of the smooth projective surface X and let C be a connected curve such that $p^{-1}(C)$ has no connected component all of whose irreducible components are compact. Let E be the union of those irreducible components D of C such that $p^{-1}(D)$ has only compact irreducible components. Let $\{a_\nu\}_{\nu=1,2,\ldots}$ be any sequence of numbers tending sufficiently rapidly to ∞ (more precisely: $a_{\nu+1} - a_\nu$ is sufficiently large).

Then, there exists a neighbourhood U of C in X and neighbourhoods V, W of E ($V \subset W \subset U$) and a continuous plurisubharmonic function φ on $p^{-1}(U)$ such that:

1) $\{x \in p^{-1}(U) \mid \varphi(x) < c\} \subset\subset \widehat{X}$ for every $c \in \mathbf{R}$;

---

* Napier's proof of this result has been written up in full in his doctoral dissertation which is being submitted to the *Mathematische Annalen*.

2) for every $\nu \geq 1$ the function φ is strongly plurisubharmonic on (a neighbourhood of) the set $\{x \in p^{-1}(U) \mid \varphi(x) = a_\nu\}$;
3) $\varphi(p^{-1}(W)) \cap \{a_\nu\}_{\nu \geq 1} = \phi$ and φ is locally constant on $p^{-1}(V)$;
4) φ is strongly plurisubharmonic on $p^{-1}(U \setminus W)$.

It is easily seen that it suffices to prove this result after blowing up points on X finitely many times. It can therefore be assumed that C has the form

$$C = A_1 \cup \cdots \cup A_k \cup E_1 \cup \cdots \cup E_l,$$

where the A_j are pairwise disjoint smooth curves such that no connected component of $p^{-1}(A_j)$ is compact, while the curves E_i (possibly singular) are pairwise disjoint and $p^{-1}(E_i)$ has only compact connected components; moreover, each A_j meets at least one of the E_i and all such intersections occur transversally at smooth points of E_i.

Let $p^{-1}(E_i) = \bigcup_{\nu \geq 1} E_i^\nu$ be the decomposition into (compact) connected components. Let $E^\nu = \bigcup_{i=1}^{l} E_i^\nu$. The most difficult point of the proof of the result stated above consists in the following:

Let $\{t_\nu\}_{\nu \geq 1}$ be a sequence of real numbers, $t_\nu \to \infty$. Let $x_0 \in C \setminus \bigcup_{i=1}^{l} E_i$. There exist neighbourhoods V of C, W of E (depending on x_0) and a neighbourhood Ω of x_0 such that $W \subset V$, $\Omega \subset V$ and there exists a continuous plurisubharmonic function ψ on $p^{-1}(V)$ such that

(a) ψ maps $p^{-1}(W)$ into $\{t_\nu\}_{\nu \geq 1}$;
(b) $\{x \in p^{-1}(W) \cup p^{-1}(\Omega) \mid \psi(x) < c\} \subset\subset \widehat{X}$ for every $c \in \mathbf{R}$;
(c) $\psi(E^\nu) = \{t_\nu\}$ for all $\nu \geq 1$.

The main ingredients in the proof are Siu's theorem about the existence of Stein neighbourhoods for a Stein subvariety of any analytic space and the fact that on an open Riemann surface one can approximate *simultaneously* holomorphic functions given on a locally finite, pairwise disjoint sequence of discs on the surface.

Full details of these constructions will be found in [7].

The condition that $\widetilde{M}$ contains no connected non-compact curve C all of whose irreducible components are compact is clearly necessary: any holomorphic function on $\widetilde{M}$ is constant on C. Thus, Theorem 2.3 shows that for smooth projective surfaces M with $b_1(M) > 0$ the Shafarevich conjecture is equivalent to the following:

Let C be a connected curve on M, and $C = \bigcup C_j$ its decomposition into irreducible components. Assume that the image of the natural map $\pi_1(C_j) \to \pi_1(M)$ is *finite* for each j; then the image of $\pi_1(C)$ in $\pi_1(M)$ is finite.

It turns out (from Napier's proof of Theorem 2.3) that this last statement needs to be checked only for certain special curves C which either occur as the full fibre of a holomorphic map of M onto a smooth curve, or which can be blown down to a point, so that, in either case, the self-intersection C^2 is non-positive. Thus, it would seem that the Shafarevich conjecture leads to a statement which is "dual" to a remarkable conjecture made by M.V. Nori [9]:

NORI'S CONJECTURE. Let $D = \sum n_j C_j$ be an effective divisor on the smooth projective surface M (the C_j are irreducible curves). Assume that $D^2 > 0$ and let $\widetilde{C}_j$ be the normalization of C_j. Then the normal subgroup of $\pi_1(M)$ generated by the images of the $\pi_1(\widetilde{C}_j)$ has finite index in $\pi_1(M)$.

In view of Theorem 2.3, one has to consider, for the Shafarevich conjecture, smooth projective surfaces M for which $\pi_1(M)$ is infinite but $b_1(M) = 0$. One way to construct such surfaces is the following.

Let D be an irreducible hermitian symmetric domain in $\mathbf{C}^n$, and let Γ be a discrete torsion free group of analytic automorphisms of D such that $X = D/\Gamma$ is compact. Then, if the rank of D is greater than 1, we have $b_1(X) = 0$ (by a theorem of Matsushima [6]). Let Y be any projective manifold with $b_1(Y) = 0$. Then, by imbedding $X \times Y$ in $\mathbf{P}^N$ and taking successive generic hyperplane sections, one obtains, because of the Lefschetz theorem, a smooth projective surface M with $\pi_1(M) \cong \pi_1(X) \times \pi_1(Y)$, which then has the required properties.

This leads to the following

QUESTION. *Let M be a smooth projective surface with $\pi_1(M)$ infinite and $b_1(M) = 0$. Does M admit a non-constant holomorphic map into a compact quotient of a bounded domain by a discrete group of automorphisms?*

For surfaces M which admit such holomorphic maps, the universal covering $\widetilde{M}$ is again holomorph-convex if it does not contain connected non-compact curves with only compact irreducible components.

One way to approach this question might be to ask when a projective surface parametrizes a non-trivial complex variation of Hodge structures. It is along these lines that Carlos Simpson has characterized compact Kähler manifolds which are uniformized by hermitian symmetric domains. Moreover, work of Hitchin and Kevin Corlette shows that it would be sufficient to construct rigid finite dimensional representations of $\pi_1(M)$. Thus, one is led to the following question:

Let M be a smooth projective surface, and let $\Gamma = \pi_1(M)$. Assume that the commutator subgroup $[\Gamma, \Gamma]$ has finite index in Γ. Does there exist a non-trivial representation $\rho\colon \Gamma \to \mathbf{GL}(V)$ on a finite dimensional vector space V such that any representation $\rho'\colon \Gamma \to \mathbf{GL}(V)$ sufficiently close to ρ is, in fact, equivalent to ρ?

Very little is known at present about these questions, or about Nori's conjecture and its "dual".

References

[1] A. Andreotti, E. Vesentini: *Carleman estimates for the Laplace-Beltrami equation on complex manifolds.* Pubbl. Math. Inst. Hautes Études Scient. **25** (1965), 81–130.

[2] G. Elencwajg: *Pseudoconvexité locale dans les variétés kählériennes.* Annales Inst. Fourier **25** (1975), 295–314.

[3] H. Grauert: *On Levi's problem and the imbedding of real analytic manifolds.* Annals of Math. **68** (1958), 460–472.

[4] H. Grauert: *Bemerkenswerte pseudokonvexe Mannigfaltigkeiten.* Math. Zeit. **81** (1963), 377–391.

[5] L. Hörmander: *L^2 estimates and existence theorems for the $\overline{\partial}$ operator.* Acta Math. **113** (1965), 89–152.

[6] Y. Matsushima: *On the first Betti number of compact quotients of higher dimensional symmetric spaces.* Annals of Math. **75** (1962), 312–330.

[7] T. Napier: **Convexity properties of coverings of smooth projective varieties.** Ph.D. Thesis, University of Chicago, 1989.

[8] R. Narasimhan: *The Levi problem in the theory of functions of several complex variables.* In **Proceedings of the international congress of mathematics**, Stockholm, 1962, 385–388.

[9] M.V. Nori: *Zariski's conjecture and related problems.* Ann. Sci. École Norm. Sup. Paris **16** (1983), 305–344.

A Banach-Steinhaus Theorem for Weak* and Order Continuous Operators

by H.H. Schaefer

Mathematisches Institut der Eberhard-Karls-Universität
Auf der Morgenstelle 10, D-7400 Tübingen

Dedicated to Prof. E. Vesentini on his sixtieth birthday

Introduction

The classical theorem of Banach-Steinhaus asserts the following: if E is a linear topological Baire space, F a topological vector space, and if (T_n) is a sequence of continuous linear maps $E \to F$ such that $(T_n x)$ converges in F for each $x \in E$, then the limiting map $T: E \to F$ is continuous. The proof (see, for example, [2, §15; 13] or [6, III.4]) shows the result to be a fairly easy consequence of another theorem of Banach, to the effect that under the above assumptions on E and F every simply bounded subset $H \subset L(E, F)$ is equicontinuous; this latter fact is better known as the *principle of uniform boundedness.* A closer analysis shows this principle to rest on the fact that in a Baire space, a closed, circled and absorbing subset has non-void interior; in the case of locally convex spaces E, F it suffices that every convex, circled, closed and absorbing set be a neighborhood of zero. A locally convex space E with this property is customarily called *barreled (tonnelé);* thus if E is a barreled and F any locally convex space, the principle of uniform boundedness is valid in $\mathcal{L}(E, F)$ by *fiat.* The merit, of course, of this concept lies in the fact that the class of barreled spaces includes but is substantially larger than the class of (locally convex) Baire spaces. Suffice it to say that barreledness is stable under the formation of inductive topologies and products of arbitrary families (see, for example, [6, IV.4.3 Cor. 3]).

There are, however, situations in analysis where a theorem of Banach-Steinhaus type is desirable (and indeed valid) but where equicontinuity arguments necessarily fail. For example, if (A_n) is a sequence of matrix operators on ℓ^∞ that converges pointwise for the weak* topology, is the limiting operator A again a matrix operator? The answer is positive (see Proposition 4.1 below), but equicontinuity of (A_n) with respect to the weak* topology will not hold except in very special cases. Generally, when dealing with weak or weak* topologies, equicontinuity of a set of continuous linear maps is an excessively restrictive requirement (see Section 2 below). Therefore, the Banach-Steinhaus theorem proved in this paper (see Theorem 3.3 below and [8, Thm. B]) rests on a rather different approach. It exploits, on the one hand, the close relationship between weak* and order continuous linear maps on certain classes of Riesz spaces, notably $L^\infty(\mu)$ (see Section 1); on the other hand, it is based on the fact that in the order dual of a wide class of Riesz spaces (see Definition 3.1 below) that includes all Dedekind σ-complete spaces, bands are weak* sequentially complete [7, II.5.10]. This, in turn, results from the fact that in these spaces a sufficient number of principal ideals are Grothendieck spaces in the usual sense [5]. Thus the prime ingredients of Theorem 3.3 below are the Grothendieck property and a new dual characterization of order continuity [8]. (Perhaps it should be mentioned that a Banach-Steinhaus theorem, unrelated to order structures, is valid for sequences of weak* continuous operators on E' whenever E is a weakly sequentially complete Banach space; but this will be fairly obvious to the interested reader).

Thus the present paper is an extension and elaboration of ideas first developed in [8]. Notation and terminology generally follows [6] and [7], except that we denote the order dual of a Riesz space E by E^*.

1. Weak* and Order Continuous Operators

For arbitrary Riesz spaces E and F, we denote by E_n^* and F_n^* their respective order continuous duals and by $L_n(E,F)$ the space of all (order bounded) order continuous linear maps $E \to F$ (cf. [1], [4] and [7]). From [8] we recall the following basic result, where E^* is the (Riesz) space of all order bounded linear forms on E.

Theorem 1.1: *If E is any Riesz space and J is an ideal of E^* separating E, then the positive cone $E_+ = \{x \in E \mid x \geq 0\}$ is $\sigma(E,J)$-closed.*

An elementary proof of Theorem 1.1 can be found in [8], but its validity can also be seen as follows. The topology $o(E,J)$ on E of uniform convergence on the order intervals of J (sometimes called the *weak Riesz topology* of E with respect to J) is a locally convex Hausdorff topology for which the lattice operations are continuous, and which is consistent with the duality $\langle E,J\rangle$ because order intervals of J are $\sigma(E^*,E)$-compact. Thus E_+ is $o(E,J)$-closed and hence, being convex, $\sigma(E,J)$-closed. |

As a first application, we note the following consequence of Theorem 1.1 which is related to the classical convergence theorem of U. Dini.

Proposition 1.2: *Let A be a directed ($\leq$) subset of E which converges to $a \in E$ for $\sigma(E,J)$, J being an ideal of E^* that separates E. Then A converges for $\sigma(E,E_n^*)$; in particular, if E is a Banach lattice with order continuous norm, then $\lim A = a$ for the norm topology.*

Proof: In fact, since E_+ is $\sigma(E,J)$-closed by Theorem 1.1, we have $a = \sup A$ [7, II.5.8] whence it follows that $a = \lim A$ for $\sigma(E,E_n^*)$. The second assertion now follows from [7, II.5.9 and II.5.10]. |

For another consequence of Theorem 1.1 that will be needed below we recall that a linear operator $T\colon E \to F$ is called *regular* if T is the difference of two positive operators. The linear space $L^r(E,F)$ of all regular operators is a Dedekind complete Riesz space (under its natural ordering) whenever F is Dedekind complete [1], [7]; in this case, T is regular iff it is order bounded. As usual, T^* denotes the algebraic adjoint of T.

Proposition 1.3: *Let E and F be Riesz spaces, with F Dedekind complete and separated by F_n^*. For an operator $T \in L^r(E,F)$, the following assertions are equivalent:*

(a) *T is order continuous;*

(b) *T is $o(E,E_n^*) - o(F,F_n^*)$ continuous;*

(c) *$T^*(F_n^*) \subset E_n^*$.*

For the proof, we refer to [8, Prop. 2].

It follows from elementary considerations that whenever $T: E \to F$ is a regular linear operator, its adjoint $T^*: F^* \to E^*$ (and, in particular, the restriction of T^* to F_n^*) is order continuous; in fact, this is so because T^* is weak* continuous (cf. proof of Proposition 1.2 above). But when is the converse true, i.e., when do the weak* continuous operators $F_n^* \to E^*$ agree with the order continuous operators? It turns out that this occurs exactly when the range space F is perfect. Perfect Riesz spaces [1], [3] are the precise generalizations of Köthe's *vollkommene Folgenräume*, cf. [2]. We recall their definition [1].

Definition 1.4: A Riesz space E is said *perfect* if the evaluation map $E \to (E_n^*)_n^*$ is a (surjective) Riesz isomorphism.

Note that this definition implies E to be Dedekind complete and separated by E_n^*; if E is any Riesz space, the Riesz spaces E^* and E_n^* are perfect [1, 33 D]. A well known special class of perfect spaces is the class of weakly sequentially complete Banach lattices (KB-spaces; cf. [7, II.10.6]); this shows that not all perfect spaces are dual Riesz spaces. We now obtain this characterization.

Theorem 1.5: *Let F be a Dedekind complete Riesz space separated by F_n^*. These assertions are pairwise equivalent:*

(a) *F is perfect.*

(b) *F is complete for the weak Riesz topology $o(F,F_n^*)$.*

(c) *A linear form f on F_n^* is weak* (i.e., $\sigma(F_n^*,F)$-) continuous whenever f is weak* continuous on order bounded subsets of F_n^*.*

(d) *For any Riesz space E, a regular linear operator $T: F_n^* \to E^*$ is weak* continuous iff T is order continuous.*

Proof: (a)$\Longrightarrow$(b): The completion $\widetilde{F}$ of F with respect to $o(F,F_n^*)$ can be identified with the space of those linear forms f on F_n^* which are $\sigma(F_n^*,F)$-continuous on each order interval of F_n^* [6, IV.6.2]. But clearly, each such f is order continuous on F_n^*, hence we have $f \in F$ and thus $F = \widetilde{F}$.

(b)$\Longrightarrow$(a): By a basic theorem of Riesz theory ([1, 32 C] or [7, II.4.12]) the evaluation map $F \to (F_n^*)_n^*$ is injective and onto an order dense ideal; thus each

$0 \leq f \in (F_n^*)_n^*$ is the pointwise (on F_n^*) limit of an increasing (directed $\leq$) family (f_α) in F_+. Since such a family is necessarily a Cauchy family for $o(F, F_n^*)$, it follows that $f \in F$; hence F is perfect.

(b)$\Longleftrightarrow$(c) is just Grothendieck's completeness theorem [6, IV.6.2].

(a)$\Longrightarrow$(d): It was observed that weak* continuity of T implies order continuity. Conversely, suppose $T: F_n^* \to E$ is order continuous. If now $\widetilde{E}$, $\widetilde{F}$ denote the order continuous duals of E^*, F_n^*, respectively, it follows that $T^*(\widetilde{E}) \subset \widetilde{F}$. But $F = \widetilde{F}$ by assumption; whence it follows that $T^*(E) \subset F$, i.e., that T is weak* continuous.

(d)$\Longrightarrow$(a): Let f be an order continuous linear form on F_n^*; for $E^* = \mathbb{R}$, f is an order continuous operator $F_n^* \to \mathbb{R}$. By hypothesis, $f \in F$ whence $(F_n^*)_n^* = F$, i.e., F is perfect. |

Let us note this classical example, where μ is a (positive) measure defined on a σ-finite (or at least localizable) measure space, so that $L^\infty(\mu) = L^1(\mu)'$.

Corollary 1.6: *Let T denote a linear, regular (equivalently, order bounded) operator from $L^\infty(\mu)$ into a dual Riesz space E^* (such as $L^q(\mu)$, $1 < q \leq +\infty$). Then T is weak* continuous if and only if T is order continuous.*

The result follows from Theorem 1.5 by observing that $F = L^1(\mu)$ is perfect.

2. Equicontinuity for the Weak Riesz Topology

We are interested in the question if a theorem of Banach-Steinhaus type holds for order continuous operators on suitable classes of Riesz spaces; as shown by Theorem 1.5 and Corollary 1.6 of the preceding section, any possible result will apply as well to weak* continuous operators on a large class of dual Riesz spaces. As explained in the Introduction, the classical Banach-Steinhaus theorem for sequences of continuous linear operators between Banach spaces rests on the equicontinuity (or uniform boundedness) principle. However, this principle can rarely be applied to weak or weak* topologies; in fact, the only locally convex spaces which are barreled for the weak topology are the spaces $\mathbb{R}^A$ or $\mathbb{C}^A$ and certain subspaces (A is any index set). Generally, if E and F are locally convex spaces, equicontinuity of a family $H \subset L(E, F)$ for the weak topologies means that the family H^* of adjoints maps every finite dimensional subspace of F' into a (common) finite dimensional subspace of E' in a uniformly bounded manner; clearly, this is too severe a restriction for interesting applications.

In this section, we shall briefly investigate the equicontinuity concept for families of order continuous operators between Riesz spaces E and F; in view of Proposition 1.3, the appropriate topologies for this undertaking are the weak Riesz topologies $o(E, E_n^*)$ and $o(F, F_n^*)$. Recall that $L_n(E, F)$ stands for the space of all (order bounded and) order continuous linear maps from E into F.

Proposition 2.1: *Let E, F denote Riesz spaces with F Dedekind complete and separated by F_n^*. A family $H \subset L_n(E, F)$ is equicontinuous for the topologies*

$o(E, E_n^*)$ and $o(F, F_n^*)$ if and only if for every $\varphi \in F_n^*$ the set $\{T^*\varphi \mid T \in H\}$ is order bounded in E_n^*.

Proof: If φ and ψ run through the respective positive cones of F_n^* and E_n^*, the polars $V = [-\varphi, \varphi]^\circ \subset F$ and $U = [-\psi, \psi]^\circ \subset E$ constitute 0-neighborhood bases for $o(F, F_n^*)$ and $o(E, E_n^*)$, respectively. Moreover, by Proposition 1.3 the order continuity of each $T \in H$ is equivalent to the continuity of T for these topologies. Thus by well known facts of operator and duality theory (see [6, III.4.1 and IV.2.3]), equicontinuity of H is equivalent to requiring that for each $\varphi \in (F_n^*)_+$ there exists $\psi \in (E_n^*)_+$ such that $H^*([-\varphi, \varphi]) \subset [-\psi, \psi]$, i.e., that $H^*([-\varphi, \varphi])$ be order bounded in E_n^*.

On the other hand, since F is Dedekind complete, $L_n(E, F)$ is a Riesz space and from standard formulas for the modulus $|T|$ of T (cf. [7, p. 229, (2)]) it follows (with the above notation) that $T(U) \subset V$ iff $|T|(U) \subset V$. Thus equicontinuity of H is equivalent to the order boundedness of $\{T^*\varphi \mid T \in H\}$ for each $\varphi \in F_n^*$. |

From the preceding proposition it is now easy to obtain a Banach-Steinhaus type result based on an equicontinuity principle. However, for all its generality, it is a meager result because the restriction placed on H is still too severe for many applications. Thus in the subsequent section, we shall look for a different approach.

Corollary 2.2: *Let E, F be Riesz spaces as in Proposition 2.1, and $\{T_\alpha \mid \alpha \in A\}$ a directed family of order continuous operators $E \to F$. If for each $\varphi \in F_n^*$ the set $\{T_\alpha^*\varphi \mid \alpha \in A\}$ is order bounded in E_n^* and if for all $x \in E$ $\lim_\alpha T_\alpha x =: Tx$ exists in F for $\sigma(F, F_n^*)$, then $T: E \to F$ is order continuous.*

Proof: By a general result on equicontinuous sets of linear mappings [6, III.4.3], the closure of the set $\{T_\alpha \mid \alpha \in A\}$ in $L(E, F)$ (for the topology of simple convergence where F carries the weak topology $\sigma(F, F_n^*)$), is equicontinuous. Thus T is continuous for the topologies $o(E, E_n^*)$ and $o(F, F_n^*)$ and hence, by Proposition 1.3, order continuous. |

3. A Banach-Steinhaus Theorem

A Banach space E is called a *Grothendieck space* if every weak* convergent sequence in E' converges weakly (i.e., for $\sigma(E', E'')$). The first non-trivial (that is, non-reflexive) example of a such a space was given by Grothendieck (1953): every Dedekind complete space $C(K)$ (K compact) has the said property. Other examples are the spaces $C(K)$ where K is quasi-Stonian (equivalently, $C(K)$ Dedekind σ-complete) (Ando 1961) or an F space (Seever 1968); still other examples are the Baire classes B_α, $\alpha \geq 1$, over a compact space (Dashiell 1981) and certain classes of Banach lattices discovered by Räbiger [5]. (We refer to [5] for an excellent survey of the history of the subject and many interesting results concerning the relation of Grothendieck spaces and Banach lattices).

Let E be a Riesz space. If (and only if) E is relatively uniformly complete [4], each principal ideal E_x ($x \geq 0$) is a Banach space under the norm whose closed unit ball is the order interval $[-x, x]$, and isometrically isomorphic to some $C(K)$ (Kakutani-Krein).

Definition 3.1: A Riesz space E is said to be of *Grothendieck type* (or an *RG-space*) if for all x in a cofinal subset of E_+, the principal ideal E_x is a Grothendieck space.

Thus, in particular, each Dedekind σ-complete Riesz space is an *RG*-space. The following lemma is now the key for the Banach-Steinhaus theorem we are going to prove.

Lemma 3.2: *If E is an RG-space, every band in the order dual E^* of E is $\sigma(E^*, E)$-sequentially complete.*

The proof is identical to the proof (due to the author) of [7, II.10.5], where the Dedekind σ-completeness of E was used only to infer that E is an *RG*-space. Therefore, the somewhat technical proof of [7, II.10.5] will not be repeated here.

Theorem 3.3: *Let E be an RG-space, and let F be a Riesz space separated by F_n^*.*
(i) *If (T_n) is a sequence of positive, order continuous operators $E \to F$ such that for each $x \in E$, $(T_n x)$ converges in F for $\sigma(F, F_n^*)$, then $T: x \mapsto \lim_n T_n x$ is order continuous.*
(ii) *If, in addition, F is Dedekind complete and (T_n) is a sequence of order continuous operators converging pointwise (for $\sigma(F, F_n^*)$) to an order bounded operator $T: E \to F$, then T is order continuous.*

Proof: (i) Let $0 \leq \varphi \in F_n^*$ be arbitrary; then $(T_n^* \varphi)$ is a $\sigma(E^*, E)$-convergent sequence contained in E_n^*. By Lemma 3.2, $T^* \varphi =: \lim_n T_n^* \varphi$ is in E_n^*; whence it follows that $T^*(F_n^*) \subset E_n^*$. Since T is positive (note that, by Theorem 1.1, the positive cone F_+ is $\sigma(F, F_n^*)$-closed), Proposition 1.3 implies that T is order continuous.

(ii) The preceding argument is again applicable, since T is order bounded, and hence Proposition 1.3 is valid for T. |

Simple examples (such as $E = F = \ell^2$) show that in assertion (ii) above, the hypothesis that T be order bounded cannot be omitted.

4. Examples.

In conclusion, we want to illustrate the usefulness of Theorem 3.3 above by two examples.

First, we consider the Riesz space ℓ^∞. A linear operator $A: \ell^\infty \to \ell^\infty$ is called a *matrix operator* if $(Ax)_i = \sum_{k=1}^{\infty} \alpha_{ik} \xi_k$ for each $i \in \mathbb{N}$, $x = (\xi_k) \in \ell^\infty$ and suitable sequences $(\alpha_{ik})_{k \in \mathbb{N}}$ (necessarily contained in ℓ^1). It is well known that an endomorphism A of ℓ^∞ is a matrix operator if and only if A is order continuous.

Proposition 4.1: *Let (A_n) be a sequence of matrix operators on ℓ^∞. If*

$$\lim_n \langle A_n x, z\rangle = \langle Ax, z\rangle$$

exists for all pairs $(x,z) \in \ell^\infty \times \ell^1$, then A is a matrix operator whose entries are the limits, as $n \to \infty$, of the respective entries of A_n.

Proof: Since for each $x \in \ell^\infty$ the sequence $(A_n x)$ is weak* and hence norm bounded, it follows that A is a norm bounded operator on ℓ^∞. But, as is well known, every norm bounded operator on ℓ^∞ is order bounded; thus the assertion follows from Theorem 3.3.(ii). (Noting that, by Theorem 1.5, weak* continuity of an endomorphism of ℓ^∞ is equivalent to order continuity, the result also follows using the weak sequential completeness of ℓ^1). |

Our second example is somewhat more delicate. Let F be a weakly sequentially complete Banach lattice; then F is a perfect Riesz space (Definition 1.4) whose order continuous dual F_n^* agrees with its Banach (and order) dual F'. Suppose $(T_t)_{t\in\mathbb{R}^+}$ is a one-parameter semigroup of positive linear operators on F' which is weak* operator continuous, i.e., such that for all pairs $(f,g) \in F \times F'$ the mapping $t \mapsto \langle T_t g, f\rangle$ is continuous on $\mathbb{R}_+$. Then it is somewhat tedious but not difficult to verify that (T_t) satisfies estimates $\|T_t\| \le Me^{\omega t}$ for suitable $M, \omega \in \mathbb{R}$, and that for g in a weak* dense linear subspace $D(Z)$, $Zg =: \lim_{t\to 0} t^{-1}(T_t g - g)$ exists for $\sigma(F', F)$. Z is a weak* closed linear operator whose spectrum $\sigma(Z)$ lies in a left half-plane $\{\lambda \in \mathbb{C} \mid \operatorname{Re}\lambda \le s(Z)\}$; for $\operatorname{Re}\lambda > s(Z)$, the resolvent $(\lambda - Z)^{-1}$ exists (as a bounded linear operator on F') and it is positive for $\lambda \in \mathbb{R}$, $\lambda > s(Z)$. We claim:

Proposition 4.2: *In addition to the preceding assumptions suppose that for some $\lambda \in \mathbb{R}$, $\lambda > s(Z)$, the resolvent $(\lambda - Z)^{-1}$ is order continuous. Then each operator T_t $(t \in \mathbb{R}_+)$ is weak* continuous, and we have $T_t = S_t'$ $(t \in \mathbb{R}_+)$ for a (unique) C_0-semigroup (S_t) on F.*

Proof: By Theorem 1.5, weak* and order continuity are equivalent properties for order bounded operators on F' $(= F_n^*)$. Moreover, by Theorem 3.3.(i) the cone of positive, order continuous operators on F' is complete under the operator norm; therefore, analytic continuation shows that $(\lambda - Z)^{-1}$ is order continuous for all λ, $\lambda > s(Z)$. Now the well known inversion formula

$$T_t = \lim_n \left(1 - \frac{t}{n} Z\right)^{-n}$$

holds for the weak* operator topology on F', by virtue of the continuity of $t \mapsto T_t$. Thus from Theorem 3.3.(i) it follows that each T_t is order and hence $\sigma(F', F)$-continuous. Thus $T_t = S_t'$ $(t \in \mathbb{R}_+)$ for a unique positive semigroup (S_t) on F which is obviously weakly continuous. But this is well known to imply the continuity of (S_t) for the strong operator topology; consequently, (S_t) is a C_0-semigroup on F. |

The preceding result was obtained in [8] for the classical case $F = L^1(\mu)$, $F' = F_n^* = L^\infty(\mu)$.

References.

[1] Fremlin, D.H.: **Topological Riesz Spaces and Measure Theory.** Cambridge University Press, Cambridge, 1974.

[2] Köthe, G.: **Topological Vector Spaces, I.** Springer, New York, 1969.

[3] Luxemburg, W.A.J. and Zaanen, A.C.: *Notes on Banach Function Spaces, VI–VII.* Proc. Nederl. Acad. Wetensch. (A) **66** (1963), 655–681.

[4] Luxemburg, W.A.J. and Zaanen, A.C.: **Riesz Spaces, I.** North Holland, Amsterdam-London, 1971.

[5] Räbiger, F.: *Beiträge zur Strukturtheorie der Grothendieck-Räume.* Sitz. Ber. Heidelberger Akad. Wiss. Nr. 4 (1985), Springer.

[6] Schaefer, H.H.: **Topological Vector Spaces.** GTM 3, 5th printing, Springer, NewYork, 1986.

[7] Schaefer, H.H.: **Banach Lattices and Positive Operators.** Springer, New York, 1974.

[8] Schaefer, H.H.: *Dual Characterization of Order Continuity and Some Applications.* Archiv d. Math. **49** (1988).

Fixed Points of Holomorphic Mappings

by Jean-Pierre Vigué

Université de Poitiers

1. Introduction

Let D be a bounded domain in $\mathbb{C}^n$ (or, more generally, in a complex Banach space E). Let $f: D \to D$ be a holomorphic mapping. The set

$$\operatorname{Fix} f = \{x \in D \mid f(x) = x\}$$

has been studied by many people. Let us recall first the following theorem proved by E. Vesentini [8 and 9]:

Theorem 1.1: *Let B be the open unit ball of a complex Banach space E. Suppose that every point x belonging to the boundary ∂B of B is a complex extreme point of $\overline{B}$. Let $f: B \to B$ be a holomorphic mapping such that $f(0) = 0$. Then*

$$\operatorname{Fix} f = B \cap F,$$

where

$$F = \{v \in E \mid f'(0) \cdot v = v\}$$

is the eigenspace of the derivative $f'(0)$ of f at the origin for the eigenvalue 1. Moreover, if E is reflexive, there exists a projection $p: E \to F$ of norm 1. So, $B \cap F$ is the image of a linear retraction $B \to B \cap F$.

The proof is based on the notion of complex geodesic. In fact, E. Vesentini [9] proved that, given $x \in D$, there exists a unique complex geodesic through the origin and x, and, more or less, this argument concludes the proof. But, in general, complex geodesics are not unique, and Vesentini's proof cannot be generalized. For example, the case of the bidisc $\Delta \times \Delta$ has been studied by M. Hervé [6] and E. Vesentini [8], and they proved the following result:

Theorem 1.2: *Let $f: \Delta \times \Delta \to \Delta \times \Delta$ be a holomorphic mapping. The set* Fix f *is one of the following sets:*

1. *the empty set $\emptyset$;*
2. *one point;*
3. *there exists a holomorphic mapping $\varphi: \Delta \to \Delta$ such that*

$$\operatorname{Fix} f = \{(\zeta_1, \zeta_2) \in \Delta \times \Delta \mid \zeta_2 = \varphi(\zeta_1)\}$$

or

$$\operatorname{Fix} f = \{(\zeta_1, \zeta_2) \in \Delta \times \Delta \mid \zeta_1 = \varphi(\zeta_2)\};$$

4. $\Delta \times \Delta$.

So, in this example, the set $\operatorname{Fix} f$ is not a linear subspace, but it is always a connected submanifold.

Now, we are going to give the results of this talk, and, first, we will begin with the finite-dimensional case.

2. Bounded domains in $\mathbb{C}^n$

We begin with the following result:

Theorem 2.1: ([13]) *Let D be a bounded domain in $\mathbb{C}^n$ and let $f: D \to D$ be a holomorphic mapping. Then $\operatorname{Fix} f$ is a complex submanifold of D. If $a \in \operatorname{Fix} f$, its tangent space $T_a(\operatorname{Fix} f)$ is equal to*

$$F = \{v \in \mathbb{C}^n \mid f'(a) \cdot v = v\}.$$

The proof of this result uses ideas of H. Cartan [3] and E. Bedford [1]. Let $a \in \operatorname{Fix} f$, and let us consider the sequence $f^p = f \circ \cdots \circ f$ (p times) of iterates of f. We can find a sequence of integers $p_j \to +\infty$ such that $q_j = p_{j+1} - p_j$ and $r_j = p_{j+1} - 2p_j$ converge to $+\infty$ and that f^{p_j} converges to a holomorphic map F (uniformly on compact subsets of D). Now, by taking subsequences of the sequences q_j and r_j, we can suppose that

$$f^{q_j} \to \rho, \qquad f^{r_j} \to G.$$

By shrinking D if necessary, we can suppose that ρ, F and G send D to D. Then, by composition, one proves easily the following relations:

$$\rho \circ F = F \circ \rho = F, \qquad F \circ G = G \circ F = \rho, \qquad f \circ \rho = \rho \circ f.$$

We deduce that

$$\rho^2 = \rho \circ \rho = \rho \circ F \circ G = F \circ G = \rho.$$

So, ρ is a holomorphic retraction, and, by a result of H. Cartan [4], there exists a local coordinate chart u defined on a neighbourhood U of a, such that $u(a) = 0$ and that $u \circ \rho \circ u^{-1}$ is a linear projection.

We have proved that $\rho(D)$ is a submanifold of D containing $\operatorname{Fix} f$, and it is easy to prove that f is a biholomorphic automorphism of $\rho(D)$. It is clear that $\rho(D)$ is a hyperbolic manifold ([5]), and we can apply the following result of H. Cartan [2]:

Theorem 2.2: *Let X be a complex hyperbolic manifold of finite dimension n, and let a be a point of X. Let $f \in \operatorname{Aut}(X)$ be a biholomorphic automorphism of X such that $f(a) = a$. Then there exists a local coordinate chart u defined in a neighbourhood U of a such that $u(a) = 0$ and that $u \circ f \circ u^{-1}$ is a linear automorphism of $\mathbb{C}^n$.*

This theorem applied to $f|_{\rho(D)}$ proves that $\operatorname{Fix} F$ is a submanifold of D.

Of course, $\operatorname{Fix} f$ is not connected in general; for example, consider the annulus

$$A = \{\zeta \in \mathbb{C} \mid 1/2 < |\zeta| < 2\},$$

and the automorphism f of A defined by $f(\zeta) = 1/\zeta$. In fact, as proved by P. Mazet and J.-P. Vigué [7], the components of $\operatorname{Fix} f$ do not always have the same dimensions.

3. Bounded convex domains in $\mathbb{C}^n$

Now, if we suppose that D is a bounded convex domain in $\mathbb{C}^n$, I can prove that the set $\operatorname{Fix} f$ is connected ([11 and 12]). In fact, we have the more precise result:

Theorem 3.1: ([12]) *Let D be a bounded convex domain in $\mathbb{C}^n$. Let $f: D \to D$ be a holomorphic mapping and let us assume that $\operatorname{Fix} f$ is not empty. Then there exists a holomorphic retraction $\psi: D \to \operatorname{Fix} f$.*

Idea of the proof. We consider φ_n defined by

$$\varphi_n(x) = \frac{1}{n} \sum_{p=0}^{n-1} f^p(x).$$

φ_n is a holomorphic mapping from D to D, and, by Montel's theorem, we can find a subsequence φ_{n_p} converging to φ (uniformly on compact subsets of D). φ is holomorphic, and, as D is taut, φ is a holomorphic mapping from D to D.

Let $a \in \operatorname{Fix} f$. By elementary linear algebra considerations, one proves that $\varphi'(a)$ is a linear projection onto

$$F = \{v \in \mathbb{C}^n \mid f'(a) \cdot v = v\}.$$

Now, let us define

$$\psi_n = \varphi^n.$$

Using Cauchy's inequalities, one proves that ψ_n converges uniformly on compact subsets of D to a holomorphic mapping ψ such that $\psi(D) \subset \operatorname{Fix} f$, and $\psi|_{\operatorname{Fix} f} = \operatorname{id}|_{\operatorname{Fix} f}$. The theorem is proved.

4. Bounded domains in reflexive Banach spaces

The results of this section have been proved in collaboration with P. Mazet [7].

The first idea we use to generalize these results to the case of bounded domains in reflexive Banach spaces is to consider weak topology and weak limits of sequences. However, it does not seem possible to generalize the proof of Theorem 2.1 for the following reason: if f_n (respectively, g_n) weakly converges to f (respectively, g), in general, $f_n \circ g_n$ does not converge to $f \circ g$. Fortunately, it is possible to generalize the proof I gave for bounded convex domains in $\mathbb{C}^n$, and we prove the following theorem:

Theorem 4.1: *Let D be a bounded convex domain in a reflexive Banach space E. Let $a \in D$, and let $f: D \to D$ be a holomorphic mapping such that $f(a) = a$. Then the set $\operatorname{Fix} f$ is a complex direct submanifold of D, tangent in a at*

$$F = \{v \in E \mid f'(a) \cdot v = v\},$$

and there exists a holomorphic retraction $\psi: D \to \operatorname{Fix} f$.

Idea of the proof: As in the finite-dimensional case, we define

$$\varphi_n = \frac{1}{n} \sum_{p=0}^{n-1} f^p.$$

Let us consider on the set $H(D, \overline{D})$ of holomorphic functions from D to $\overline{D}$ the topology of uniform weak convergence on finite-dimensional compact subsets of D. It is more or less standard that $H(D, \overline{D})$ is compact, and so, we can find φ adherent to the sequence φ_n.

φ is a holomorphic mapping from D to D; it is clear that

$$\varphi \circ f = \varphi, \qquad \operatorname{Fix} f \subset \operatorname{Fix} \varphi,$$

and using the continuity of $f'(a)$ for the weak topology, we prove that

$$f'(a) \circ \varphi'(a) = \varphi'(a).$$

So, $\varphi'(a)$ is a projection onto F. Now, we consider the sequence of iterates

$$\psi_n = \varphi^n$$

of φ.

If g is holomorphic in a neighbourhood of a, we note

$$g = \sum_{p=0}^{\infty} P_p(g)$$

the development of g in series of homogeneous polynomials at a. We prove the following lemma.

Lemma 4.2: *For every $n \geq 0$ and $p \leq n$ we have*

$$P_p(f \circ \varphi^n) = P_p(\varphi^n) = P_p(\varphi^{n+1}).$$

We have already proved this lemma for $n = 1$, and the proof is by induction on n.

Using Cauchy's inequalities for bounded mappings, this lemma implies that ψ_n converges to a limit ψ uniformly on a ball of center a and of radius small enough. But, by [10], it implies that ψ_n converges to a limit ψ for the topology of local uniform convergence. So, $\psi \in H(D, D)$, and we have $f \circ \psi = \psi$, $\psi^2 = \psi$, $\operatorname{Fix} f = \operatorname{Fix} \psi$. As ψ is a holomorphic retraction, the theorem is a consequence of H. Cartan [4].

Now, if we do not suppose that D is convex, we can also define φ_n and φ. The only difference is that φ does not send D to D. However, if a is a fixed point of f, we prove in [7] that there exists a neighbourhood U of a such that $\varphi(U) \subset U$, and, with some small changes, we can generalize the proof to this case.

Theorem 4.3: (P. Mazet and J.-P. Vigué [7]) *Let D be a bounded domain in a reflexive Banach space E, and let $f: D \to D$ be a holomorphic mapping. Then the set* $\operatorname{Fix} f$ *is a complex direct submanifold of D.*

5. An example

To conclude this talk, I shall give an example [7] which proves that the conclusion of Theorems 4.1 and 4.3 is not true for every Banach space E.

Let $c_0(\mathbb{N})$ be the Banach space of sequences converging to 0 at infinity. Let B be the open unit ball of $c_0(\mathbb{N})$. Let $n \in \mathbb{N}$, and let f be a holomorphic mapping from the polydisk Δ^n into itself. Let us define $F: B \to B$ in the following way:

$$(Z_p)_{p \in \mathbb{N}} = F\big((z_p)_{p \in \mathbb{N}}\big),$$

where

$$\begin{aligned} (Z_0, \ldots, Z_{n-1}) &= (z_0, \ldots, z_{n-1}), \\ (Z_n, \ldots, Z_{2n-1}) &= f(z_0, \ldots, z_{n-1}), \\ Z_{2n+k} &= z_{n+k}, \quad \forall k \geq 0. \end{aligned}$$

It is easy to check that

$$\operatorname{Fix} F = Z(f) \times \{0\},$$

where

$$Z(f) = \big\{(z_0, \ldots, z_{n-1}) \in \Delta^n \mid f(z_0, \ldots, z_{n-1}) = 0\big\}$$

is the zero set of f. It is clear that $Z(f)$ and $\operatorname{Fix}(F)$ are not, in general, submanifolds.

There are also examples (P. Mazet and J.-P. Vigué [7]) in which E is a dual space.

References

[1] E. Bedford: *On the automorphism group of a Stein manifold.* Math. Ann. **266** (1983), 215–227.

[2] H. Cartan: *Les fonctions de deux variables complexes et le problème de la représentation analytique.* J. Math. Pures Appl. **11** (1931), 1–114.

[3] H. Cartan: *Sur les fonctions de plusieurs variables complexes. L'itération des transformations intérieures d'un domaine borné.* Math. Z. **35** (1932), 760–773.

[4] H. Cartan: *Sur les rétractions d'une variété.* C.R. Acad. Sc. Paris **303** (1986), 715–716.

[5] T. Franzoni, E. Vesentini: **Holomorphic maps and invariant distances.** Mathematical Studies 40, North-Holland, Amsterdam, 1980.

[6] M. Hervé: *Quelques propriétés des applications analytiques d'une boule à m dimensions dans elle-même.* J. Math. Pures Appl. **42** (1963), 117–147.

[7] P. Mazet, J.-P. Vigué: *Points fixes d'une application holomorphe d'un domaine borné dans lui-même.* To appear.

[8] E. Vesentini: *Complex geodesics.* Comp. Math. **44** (1981), 375–394.

[9] E. Vesentini: *Complex geodesics and holomorphic maps.* Symp. Math. **26** (1982), 211–230.

[10] J.-P. Vigué: *Le groupe des automorphismes analytiques d'un domaine borné d'un espace de Banach complexe. Application aux domaines bornés symétriques.* Ann. Sc. Ec. Norm. Sup. **9** (1976), 203–282.

[11] J.-P. Vigué: *Géodésiques complexes et points fixes d'applications holomorphes.* Adv. Math. **52** (1984), 241–247.

[12] J.-P. Vigué: *Points fixes d'applications holomorphes dans un domaine borné convexe de $\mathbb{C}^n$.* Tr. Am. Math. Soc. **289** (1985), 345–353.

[13] J.-P. Vigué: *Sur les points fixes d'applications holomorphes.* C.R. Acad. Sc. Paris **303** (1986), 927–930.

List of Participants

Marco Abate — Scuola Normale Superiore, Pisa
Giuseppe Accascina — Università di Pisa
Francesca Acquistapace — Università di Pisa
Antonio Ambrosetti — Scuola Normale Superiore, Pisa
Vincenzo Ancona — Università di Firenze
Michael F. Atiyah — Oxford University
Bernard H. Aupetit — Université Laval
Stefan A. Bauer — Universität Göttingen
Bruno Bigolin — Università Cattolica del S. Cuore, Brescia
Fabrizio Broglia — Università di Pisa
Claudio Buzzanca — Università di Palermo
Eugenio Calabi — University of Pennsylvania
Leonardo Cangeliti — Università di Roma
Mario Carlotti — Università di Pisa
Fabrizio Catanese — Università di Pisa
Franco Cazzaniga — Università di Milano
Mauro Chiaretti — Università di Roma
Alberto Conte — Comitato per la Matematica del CNR, Roma
Paolo Cragnolini — Università di Pisa
Paolo de Bartolomeis — Università di Firenze
Maria Dedò — Università di Cagliari
Gilberto Dini — Università di Firenze
Luigia Diterlizzi — Università di Bari
James Eeels — University of Warwick
Sandro Faedo — Università di Pisa
Franco Favilli — Università di Pisa
Massimo Ferrarotti — Università di Pisa
Elisabetta Fortuna — Università di Pisa
Tullio Franzoni — Università di Pisa
Margherita Galbiati — Università di Pisa
Laura Geatti — IEI, Consiglio Nazionale delle Ricerche, Pisa
Graziano Gentili — SISSA, Trieste
Francesco Gherardelli — Università di Firenze
Giuliana Gigante — Università di Parma
Michel Goze — Université de Haute-Alsace, Mulhouse
Hans Grauert — Universität Göttingen
Renata Grimaldi — Università di Palermo
Francesco Guaraldo — Università di Roma
Shoshichi Kobayashi — University of California, Berkeley
Jerzy J. Konderak — International Centre for Theoretical Physics, Trieste
Adam Korányi — Lehman College, City University of New York

Sandro Levi — Università di Milano
Patrizia Macrì — Università di Roma
Abnedacer Makhlouf — Université de Haute-Alsace, Mulhouse
Marco Manetti — Scuola Normale Superiore, Pisa
Stefano Marchiafava — Università di Roma
Ermanno Marchionna — Università di Milano
Cesarina Marchionna Tibiletti — Università di Milano
Enzo Martinelli — Università di Roma
Francesca Menozzi — Università di Pisa
Mauro Meschiari — Università di Modena
Giuseppe Modica — Università di Firenze
Flavio Mosca — Università di Pisa
Venkatesha M.K. Murthy — Università di Pisa
Emilio Musso — Università de L'Aquila
Mauro Nacinovich — Università di Pisa
Rainer Nagel — Universität Tübingen
Antonella Nannicini — Università di Firenze
Raghavan Narasimhan — University of Chicago
Michael A. O'Connor — IBM, Yorktown Heights, NY
Paolo Oliverio — Università della Calabria
Anna Maria Pastore — Università di Bari
I.I. Piatetski-Shapiro — Tel Aviv University
Paolo Piccinni — Università di Salerno
Fabio Podestà — Scuola Normale Superiore, Pisa
Carlo Pucci — Università di Firenze
Giuseppe Puglisi — Università di Pisa
Luigi A. Radicati di Brozolo — Scuola Normale Superiore, Pisa
Fulvio Ricci — Politecnico di Torino
Hugo Rossi — University of Utah
Delfina Roux — Università di Milano
Simon M. Salamon — Oxford University
Mario Salvetti — Università di Pisa
Helmut H. Schaefer — Universität Tübingen
Sergio Spagnolo — Università di Pisa
David R. Speiner — UCL, Louvain-la-Neuve, Belgique
László L. Stachó — Bolyai Institute, Szeged
Francesco Succi — Università di Roma
Corrado Tanasi — Università di Palermo
Cesarina Tibiletti — Università di Milano
Alberto Tognoli — Università di Trento
Giuseppe Tomassini — Scuola Normale Superiore, Pisa
Franco Tricerri — Università di Firenze
Sergio Venturini — Scuola Normale Superiore, Pisa

Edoardo Vesentini — Scuola Normale Superiore, Pisa
Jean-Pierre Vigué — Université de Paris VI
Vinicio Villani — Università di Pisa
Georges G. Weill — Université de Tours
Gian Carlo Wick — Università di Torino
Paolo Zappa — Università di Camerino

LECTURE NOTES IN MATHEMATICS

Edited by A. Dold and B. Eckmann

Some general remarks on the publication of proceedings of congresses and symposia

Lecture Notes aim to report new developments - quickly, informally and at a high level. The following describes criteria and procedures which apply to proceedings volumes. The editors of a volume are strongly advised to inform contributors about these points at an early stage.

§1. One (or more) expert participant(s) of the meeting should act as the responsible editor(s) of the proceedings. They select the papers which are suitable (cf. §§ 2, 3) for inclusion in the proceedings, and have them individually refereed (as for a journal). It should not be assumed that the published proceedings must reflect conference events faithfully and in their entirety. Contributions to the meeting which are not included in the proceedings can be listed by title. The series editors will normally not interfere with the editing of a particular proceedings volume - except in fairly obvious cases, or on technical matters, such as described in §§ 2, 3. The names of the responsible editors appear on the title page of the volume.

§2. The proceedings should be reasonably homogeneous (concerned with a limited area). For instance, the proceedings of a congress on "Analysis" or "Mathematics in Wonderland" would normally not be sufficiently homogeneous.

One or two longer survey articles on recent developments in the field are often very useful additions to such proceedings - even if they do not correspond to actual lectures at the congress. An extensive introduction on the subject of the congress would be desirable.

§3. The contributions should be of a high mathematical standard and of current interest. Research articles should present new material and not duplicate other papers already published or due to be published. They should contain sufficient information and motivation and they should present proofs, or at least outlines of such, in sufficient detail to enable an expert to complete them. Thus resumes and mere announcements of papers appearing elsewhere cannot be included, although more detailed versions of a contribution may well be published in other places later.

Surveys, if included, should cover a sufficiently broad topic, and should in general not simply review the author's own recent research. In the case of surveys, exceptionally, proofs of results may not be necessary.

"Mathematical Reviews" and "Zentralblatt für Mathematik" require that papers in proceedings volumes carry an explicit statement that they are in final form and that no similar paper has been or is being submitted elsewhere, if these papers are to be considered for a review. Normally, papers that satisfy the criteria of the Lecture Notes in Mathematics series also satisfy this

.../...

requirement, but we would strongly recommend that the contributing authors be asked to give this guarantee explicitly at the beginning or end of their paper. There will occasionally be cases where this does not apply but where, for special reasons, the paper is still acceptable for LNM.

§4. Proceedings should appear soon after the meeeting. The publisher should, therefore, receive the complete manuscript within nine months of the date of the meeting at the latest.

§5. Plans or proposals for proceedings volumes should be sent to one of the editors of the series or to Springer-Verlag Heidelberg. They should give sufficient information on the conference or symposium, and on the proposed proceedings. In particular, they should contain a list of the expected contributions with their prospective length. Abstracts or early versions (drafts) of some of the contributions are very helpful.

§6. Lecture Notes are printed by photo-offset from camera-ready typed copy provided by the editors. For this purpose Springer-Verlag provides editors with technical instructions for the preparation of manuscripts and these should be distributed to all contributing authors. Springer-Verlag can also, on request, supply stationery on which the prescribed typing area is outlined. Some homogeneity in the presentation of the contributions is desirable.

Careful preparation of manuscripts will help keep production time short and ensure a satisfactory appearance of the finished book. The actual production of a Lecture Notes volume normally takes 6 -8 weeks.

Manuscripts should be at least 100 pages long. The final version should include a table of contents and as far as applicable a subject index.

§7. Editors receive a total of 50 free copies of their volume for distribution to the contributing authors, but no royalties. (Unfortunately, no reprints of individual contributions can be supplied.) They are entitled to purchase further copies of their book for their personal use at a discount of 33.3 %, other Springer mathematics books at a discount of 20 % directly from Springer-Verlag. Contributing authors may purchase the volume in which their article appears at a discount of 33.3 %.

Commitment to publish is made by letter of intent rather than by signing a formal contract. Springer-Verlag secures the copyright for each volume.

Vol. 1259: F. Cano Torres, Desingularization Strategies for Three-Dimensional Vector Fields. IX, 189 pages. 1987.

Vol. 1260: N.H. Pavel, Nonlinear Evolution Operators and Semigroups. VI, 285 pages. 1987.

Vol. 1261: H. Abels, Finite Presentability of S-Arithmetic Groups. Compact Presentability of Solvable Groups. VI, 178 pages. 1987.

Vol. 1262: E. Hlawka (Hrsg.), Zahlentheoretische Analysis II. Seminar, 1984–86. V, 158 Seiten. 1987.

Vol. 1263: V.L. Hansen (Ed.), Differential Geometry. Proceedings, 1985. XI, 288 pages. 1987.

Vol. 1264: Wu Wen-tsün, Rational Homotopy Type. VIII, 219 pages. 1987.

Vol. 1265: W. Van Assche, Asymptotics for Orthogonal Polynomials. VI, 201 pages. 1987.

Vol. 1266: F. Ghione, C. Peskine, E. Sernesi (Eds.), Space Curves. Proceedings, 1985. VI, 272 pages. 1987.

Vol. 1267: J. Lindenstrauss, V.D. Milman (Eds.), Geometrical Aspects of Functional Analysis. Seminar. VII, 212 pages. 1987.

Vol. 1268: S.G. Krantz (Ed.), Complex Analysis. Seminar, 1986. VII, 195 pages. 1987.

Vol. 1269: M. Shiota, Nash Manifolds. VI, 223 pages. 1987.

Vol. 1270: C. Carasso, P.-A. Raviart, D. Serre (Eds.), Nonlinear Hyperbolic Problems. Proceedings, 1986. XV, 341 pages. 1987.

Vol. 1271: A.M. Cohen, W.H. Hesselink, W.L.J. van der Kallen, J.R. Strooker (Eds.), Algebraic Groups Utrecht 1986. Proceedings. XII, 284 pages. 1987.

Vol. 1272: M.S. Livšic, L.L. Waksman, Commuting Nonselfadjoint Operators in Hilbert Space. III, 115 pages. 1987.

Vol. 1273: G.-M. Greuel, G. Trautmann (Eds.), Singularities, Representation of Algebras, and Vector Bundles. Proceedings, 1985. XIV, 383 pages. 1987.

Vol. 1274: N. C. Phillips, Equivariant K-Theory and Freeness of Group Actions on C*-Algebras. VIII, 371 pages. 1987.

Vol. 1275: C.A. Berenstein (Ed.), Complex Analysis I. Proceedings, 1985–86. XV, 331 pages. 1987.

Vol. 1276: C.A. Berenstein (Ed.), Complex Analysis II. Proceedings, 1985–86. IX, 320 pages. 1987.

Vol. 1277: C.A. Berenstein (Ed.), Complex Analysis III. Proceedings, 1985–86. X, 350 pages. 1987.

Vol. 1278: S.S. Koh (Ed.), Invariant Theory. Proceedings, 1985. V, 102 pages. 1987.

Vol. 1279: D. Ieşan, Saint-Venant's Problem. VIII, 162 Seiten. 1987.

Vol. 1280: E. Neher, Jordan Triple Systems by the Grid Approach. XII, 193 pages. 1987.

Vol. 1281: O.H. Kegel, F. Menegazzo, G. Zacher (Eds.), Group Theory. Proceedings, 1986. VII, 179 pages. 1987.

Vol. 1282: D.E. Handelman, Positive Polynomials, Convex Integral Polytopes, and a Random Walk Problem. XI, 136 pages. 1987.

Vol. 1283: S. Mardešić, J. Segal (Eds.), Geometric Topology and Shape Theory. Proceedings, 1986. V, 261 pages. 1987.

Vol. 1284: B.H. Matzat, Konstruktive Galoistheorie. X, 286 pages. 1987.

Vol. 1285: I.W. Knowles, Y. Saitō (Eds.), Differential Equations and Mathematical Physics. Proceedings, 1986. XVI, 499 pages. 1987.

Vol. 1286: H.R. Miller, D.C. Ravenel (Eds.), Algebraic Topology. Proceedings, 1986. VII, 341 pages. 1987.

Vol. 1287: E.B. Saff (Ed.), Approximation Theory, Tampa. Proceedings, 1985–1986. V, 228 pages. 1987.

Vol. 1288: Yu. L. Rodin, Generalized Analytic Functions on Riemann Surfaces. V, 128 pages, 1987.

Vol. 1289: Yu. I. Manin (Ed.), K-Theory, Arithmetic and Geometry. Seminar, 1984–1986. V, 399 pages. 1987.

Vol. 1290: G. Wüstholz (Ed.), Diophantine Approximation and Transcendence Theory. Seminar, 1985. V, 243 pages. 1987.

Vol. 1291: C. Mœglin, M.-F. Vignéras, J.-L. Waldspurger, Correspondances de Howe sur un Corps p-adique. VII, 163 pages. 1987

Vol. 1292: J.T. Baldwin (Ed.), Classification Theory. Proceedings, 1985. VI, 500 pages. 1987.

Vol. 1293: W. Ebeling, The Monodromy Groups of Isolated Singularities of Complete Intersections. XIV, 153 pages. 1987.

Vol. 1294: M. Queffélec, Substitution Dynamical Systems – Spectral Analysis. XIII, 240 pages. 1987.

Vol. 1295: P. Lelong, P. Dolbeault, H. Skoda (Réd.), Séminaire d'Analyse P. Lelong – P. Dolbeault – H. Skoda. Seminar, 1985/1986. VII, 283 pages. 1987.

Vol. 1296: M.-P. Malliavin (Ed.), Séminaire d'Algèbre Paul Dubreil et Marie-Paule Malliavin. Proceedings, 1986. IV, 324 pages. 1987.

Vol. 1297: Zhu Y.-l., Guo B.-y. (Eds.), Numerical Methods for Partial Differential Equations. Proceedings. XI, 244 pages. 1987.

Vol. 1298: J. Aguadé, R. Kane (Eds.), Algebraic Topology, Barcelona 1986. Proceedings. X, 255 pages. 1987.

Vol. 1299: S. Watanabe, Yu.V. Prokhorov (Eds.), Probability Theory and Mathematical Statistics. Proceedings, 1986. VIII, 589 pages. 1988.

Vol. 1300: G.B. Seligman, Constructions of Lie Algebras and their Modules. VI, 190 pages. 1988.

Vol. 1301: N. Schappacher, Periods of Hecke Characters. XV, 160 pages. 1988.

Vol. 1302: M. Cwikel, J. Peetre, Y. Sagher, H. Wallin (Eds.), Function Spaces and Applications. Proceedings, 1986. VI, 445 pages. 1988.

Vol. 1303: L. Accardi, W. von Waldenfels (Eds.), Quantum Probability and Applications III. Proceedings, 1987. VI, 373 pages. 1988.

Vol. 1304: F.Q. Gouvêa, Arithmetic of p-adic Modular Forms. VIII, 121 pages. 1988.

Vol. 1305: D.S. Lubinsky, E.B. Saff, Strong Asymptotics for Extremal Polynomials Associated with Weights on $\mathbb{R}$. VII, 153 pages. 1988.

Vol. 1306: S.S. Chern (Ed.), Partial Differential Equations. Proceedings, 1986. VI, 294 pages. 1988.

Vol. 1307: T. Murai, A Real Variable Method for the Cauchy Transform, and Analytic Capacity. VIII, 133 pages. 1988.

Vol. 1308: P. Imkeller, Two-Parameter Martingales and Their Quadratic Variation. IV, 177 pages. 1988.

Vol. 1309: B. Fiedler, Global Bifurcation of Periodic Solutions with Symmetry. VIII, 144 pages. 1988.

Vol. 1310: O.A. Laudal, G. Pfister, Local Moduli and Singularities. V, 117 pages. 1988.

Vol. 1311: A. Holme, R. Speiser (Eds.), Algebraic Geometry, Sundance 1986. Proceedings. VI, 320 pages. 1988.

Vol. 1312: N.A. Shirokov, Analytic Functions Smooth up to the Boundary. III, 213 pages. 1988.

Vol. 1313: F. Colonius, Optimal Periodic Control. VI, 177 pages. 1988.

Vol. 1314: A. Futaki, Kähler-Einstein Metrics and Integral Invariants. IV, 140 pages. 1988.

Vol. 1315: R.A. McCoy, I. Ntantu, Topological Properties of Spaces of Continuous Functions. IV, 124 pages. 1988.

Vol. 1316: H. Korezlioglu, A.S. Ustunel (Eds.), Stochastic Analysis and Related Topics. Proceedings, 1986. V, 371 pages. 1988.

Vol. 1317: J. Lindenstrauss, V.D. Milman (Eds.), Geometric Aspects of Functional Analysis. Seminar, 1986–87. VII, 289 pages. 1988.

Vol. 1318: Y. Felix (Ed.), Algebraic Topology – Rational Homotopy. Proceedings, 1986. VIII, 245 pages. 1988

Vol. 1319: M. Vuorinen, Conformal Geometry and Quasiregular Mappings. XIX, 209 pages. 1988.

Vol. 1320: H. Jürgensen, G. Lallement, H.J. Weinert (Eds.), Semigroups, Theory and Applications. Proceedings, 1986. X, 416 pages. 1988.

Vol. 1321: J. Azéma, P.A. Meyer, M. Yor (Eds.), Séminaire de Probabilités XXII. Proceedings. IV, 600 pages. 1988.

Vol. 1322: M. Métivier, S. Watanabe (Eds.), Stochastic Analysis. Proceedings, 1987. VII, 197 pages. 1988.

Vol. 1323: D.R. Anderson, H.J. Munkholm, Boundedly Controlled Topology. XII, 309 pages. 1988.

Vol. 1324: F. Cardoso, D.G. de Figueiredo, R. Iório, O. Lopes (Eds.), Partial Differential Equations. Proceedings, 1986. VIII, 433 pages. 1988.

Vol. 1325: A. Truman, I.M. Davies (Eds.), Stochastic Mechanics and Stochastic Processes. Proceedings, 1986. V, 220 pages. 1988.

Vol. 1326: P.S. Landweber (Ed.), Elliptic Curves and Modular Forms in Algebraic Topology. Proceedings, 1986. V, 224 pages. 1988.

Vol. 1327: W. Bruns, U. Vetter, Determinantal Rings. VII,236 pages. 1988.

Vol. 1328: J.L. Bueso, P. Jara, B. Torrecillas (Eds.), Ring Theory. Proceedings, 1986. IX, 331 pages. 1988.

Vol. 1329: M. Alfaro, J.S. Dehesa, F.J. Marcellan, J.L. Rubio de Francia, J. Vinuesa (Eds.): Orthogonal Polynomials and their Applications. Proceedings, 1986. XV, 334 pages. 1988.

Vol. 1330: A. Ambrosetti, F. Gori, R. Lucchetti (Eds.), Mathematical Economics. Montecatini Terme 1986. Seminar. VII, 137 pages. 1988.

Vol. 1331: R. Bamón, R. Labarca, J. Palis Jr. (Eds.), Dynamical Systems, Valparaiso 1986. Proceedings. VI, 250 pages. 1988.

Vol. 1332: E. Odell, H. Rosenthal (Eds.), Functional Analysis. Proceedings, 1986–87. V, 202 pages. 1988.

Vol. 1333: A.S. Kechris, D.A. Martin, J.R. Steel (Eds.), Cabal Seminar 81–85. Proceedings, 1981–85. V, 224 pages. 1988.

Vol. 1334: Yu.G. Borisovich, Yu. E. Gliklikh (Eds.), Global Analysis – Studies and Applications III. V, 331 pages. 1988.

Vol. 1335: F. Guillén, V. Navarro Aznar, P. Pascual-Gainza, F. Puerta, Hyperrésolutions cubiques et descente cohomologique. XII, 192 pages. 1988.

Vol. 1336: B. Helffer, Semi-Classical Analysis for the Schrödinger Operator and Applications. V, 107 pages. 1988.

Vol. 1337: E. Sernesi (Ed.), Theory of Moduli. Seminar, 1985. VIII, 232 pages. 1988.

Vol. 1338: A.B. Mingarelli, S.G. Halvorsen, Non-Oscillation Domains of Differential Equations with Two Parameters. XI, 109 pages. 1988.

Vol. 1339: T. Sunada (Ed.), Geometry and Analysis of Manifolds. Procedings, 1987. IX, 277 pages. 1988.

Vol. 1340: S. Hildebrandt, D.S. Kinderlehrer, M. Miranda (Eds.), Calculus of Variations and Partial Differential Equations. Proceedings, 1986. IX, 301 pages. 1988.

Vol. 1341: M. Dauge, Elliptic Boundary Value Problems on Corner Domains. VIII, 259 pages. 1988.

Vol. 1342: J.C. Alexander (Ed.), Dynamical Systems. Proceedings, 1986–87. VIII, 726 pages. 1988.

Vol. 1343: H. Ulrich, Fixed Point Theory of Parametrized Equivariant Maps. VII, 147 pages. 1988.

Vol. 1344: J. Král, J. Lukeš, J. Netuka, J. Veselý (Eds.), Potential Theory – Surveys and Problems. Proceedings, 1987. VIII, 271 pages. 1988.

Vol. 1345: X. Gomez-Mont, J. Seade, A. Verjovski (Eds.), Holomorphic Dynamics. Proceedings, 1986. VII, 321 pages. 1988.

Vol. 1346: O. Ya. Viro (Ed.), Topology and Geometry – Rohlin Seminar. XI, 581 pages. 1988.

Vol. 1347: C. Preston, Iterates of Piecewise Monotone Mappings on an Interval. V, 166 pages. 1988.

Vol. 1348: F. Borceux (Ed.), Categorical Algebra and its Applications. Proceedings, 1987. VIII, 375 pages. 1988.

Vol. 1349: E. Novak, Deterministic and Stochastic Error Bounds in Numerical Analysis. V, 113 pages. 1988.

Vol. 1350: U. Koschorke (Ed.), Differential Topology. Proceedings, 1987. VI, 269 pages. 1988.

Vol. 1351: I. Laine, S. Rickman, T. Sorvali, (Eds.), Complex Analysis, Joensuu 1987. Proceedings. XV, 378 pages. 1988.

Vol. 1352: L.L. Avramov, K.B. Tchakerian (Eds.), Algebra – Some Current Trends. Proceedings, 1986. IX, 240 Seiten. 1988.

Vol. 1353: R.S. Palais, Ch.-l. Terng, Critical Point Theory and Submanifold Geometry. X, 272 pages. 1988.

Vol. 1354: A. Gómez, F. Guerra, M.A. Jiménez, G. López (Eds.), Approximation and Optimization. Proceedings, 1987. VI, 280 pages. 1988.

Vol. 1355: J. Bokowski, B. Sturmfels, Computational Synthetic Geometry. V, 168 pages. 1989.

Vol. 1356: H. Volkmer, Multiparameter Eigenvalue Problems and Expansion Theorems. VI, 157 pages. 1988.

Vol. 1357: S. Hildebrandt, R. Leis (Eds.), Partial Differential Equations and Calculus of Variations. VI, 423 pages. 1988.

Vol. 1358: D. Mumford, The Red Book of Varieties and Schemes. V, 309 pages. 1988.

Vol. 1359: P. Eymard, J.-P. Pier (Eds.), Harmonic Analysis. Proceedings, 1987. VIII, 287 pages. 1988.

Vol. 1360: G. Anderson, C. Greengard (Eds.), Vortex Methods. Proceedings, 1987. V, 141 pages. 1988.

Vol. 1361: T. tom Dieck (Ed.), Algebraic Topology and Transformation Groups. Proceedings, 1987. VI, 298 pages. 1988.

Vol. 1362: P. Diaconis, D. Elworthy, H. Föllmer, E. Nelson, G.C. Papanicolaou, S.R.S. Varadhan. École d'Été de Probabilités de Saint Flour XV–XVII, 1985–87. Editor: P.L. Hennequin. V, 459 pages. 1988.

Vol. 1363: P.G. Casazza, T.J. Shura. Tsirelson's Space. VIII, 204 pages. 1988.

Vol. 1364: R.R. Phelps, Convex Functions, Monotone Operators and Differentiability. IX, 115 pages. 1989.

Vol. 1365: M. Giaquinta (Ed.), Topics in Calculus of Variations. Seminar, 1987. X, 196 pages. 1989.

Vol. 1366: N. Levitt, Grassmannians and Gauss Maps in PL-Topology. V, 203 pages. 1989.

Vol. 1367: M. Knebusch, Weakly Semialgebraic Spaces. XX, 376 pages. 1989.

Vol. 1368: R. Hübl, Traces of Differential Forms and Hochschild Homology. III, 111 pages. 1989.

Vol. 1369: B. Jiang, Ch.-K. Peng, Z. Hou (Eds.), Differential Geometry and Topology. Proceedings, 1986–87. VI, 366 pages. 1989.

Vol. 1370: G. Carlsson, R.L. Cohen, H.R. Miller, D.C. Ravenel (Eds.), Algebraic Topology. Proceedings, 1986. IX, 456 pages. 1989.

Vol. 1371: S. Glaz, Commutative Coherent Rings. XI, 347 pages. 1989.

Vol. 1372: J. Azéma, P.A. Meyer, M. Yor (Eds.), Séminaire de Probabilités XXIII. Proceedings. IV, 583 pages. 1989.

Vol. 1373: G. Benkart, J.M. Osborn (Eds.), Lie Algebras, Madison 1987. Proceedings. V, 145 pages. 1989.

Vol. 1374: R.C. Kirby, The Topology of 4-Manifolds. VI, 108 pages. 1989.

Vol. 1375: K. Kawakubo (Ed.), Transformation Groups. Proceedings, 1987. VIII, 394 pages, 1989.

Vol. 1376: J. Lindenstrauss, V.D. Milman (Eds.), Geometric Aspects of Functional Analysis. Seminar (GAFA) 1987–88. VII, 288 pages. 1989.

Vol. 1377: J.F. Pierce, Singularity Theory, Rod Theory, and Symmetry Breaking Loads. IV, 177 pages. 1989.

Vol. 1378: R.S. Rumely, Capacity Theory on Algebraic Curves. III, 437 pages. 1989.

Vol. 1379: H. Heyer (Ed.), Probability Measures on Groups IX. Proceedings, 1988. VIII, 437 pages. 1989

Vol. 1380: H.P. Schlickewei, E. Wirsing (Eds.), Number Theory, Ulm 1987. Proceedings. V, 266 pages. 1989.

Vol. 1381: J.-O. Strömberg, A. Torchinsky. Weighted Hardy Spaces. V, 193 pages. 1989.

Vol. 1382: H. Reiter, Metaplectic Groups and Segal Algebras. XI, 128 pages. 1989.

Vol. 1383: D.V. Chudnovsky, G.V. Chudnovsky, H. Cohn, M.B. Nathanson (Eds.), Number Theory, New York 1985–88. Seminar. V, 256 pages. 1989.

Vol. 1384: J. Garcia-Cuerva (Ed.), Harmonic Analysis and Partial Differential Equations. Proceedings, 1987. VII, 213 pages. 1989.

Vol. 1385: A.M. Anile, Y. Choquet-Bruhat (Eds.), Relativistic Fluid Dynamics. Seminar, 1987. V, 308 pages. 1989.

Vol. 1386: A. Bellen, C.W. Gear, E. Russo (Eds.), Numerical Methods for Ordinary Differential Equations. Proceedings, 1987. VII, 136 pages. 1989.

Vol. 1387: M. Petković, Iterative Methods for Simultaneous Inclusion of Polynomial Zeros. X, 263 pages. 1989.

Vol. 1388: J. Shinoda, T.A. Slaman, T. Tugué (Eds.), Mathematical Logic and Applications. Proceedings, 1987. V, 223 pages. 1989.

Vol. 1000: Second Edition. H. Hopf, Differential Geometry in the Large. VII, 184 pages. 1989.

Vol. 1389: E. Ballico, C. Ciliberto (Eds.), Algebraic Curves and Projective Geometry. Proceedings, 1988. V, 288 pages. 1989.

Vol. 1390: G. Da Prato, L. Tubaro (Eds.), Stochastic Partial Differential Equations and Applications II. Proceedings, 1988. VI, 258 pages. 1989.

Vol. 1391: S. Cambanis, A. Weron (Eds.), Probability Theory on Vector Spaces IV. Proceedings, 1987. VIII, 424 pages. 1989.

Vol. 1392: R. Silhol, Real Algebraic Surfaces. X, 215 pages. 1989.

Vol. 1393: N. Bouleau, D. Feyel, F. Hirsch, G. Mokobodzki (Eds.), Séminaire de Théorie du Potentiel Paris, No. 9. Proceedings. VI, 265 pages. 1989.

Vol. 1394: T.L. Gill, W.W. Zachary (Eds.), Nonlinear Semigroups, Partial Differential Equations and Attractors. Proceedings, 1987. IX, 233 pages. 1989.

Vol. 1395: K. Alladi (Ed.), Number Theory, Madras 1987. Proceedings. VII, 234 pages. 1989.

Vol. 1396: L. Accardi, W. von Waldenfels (Eds.), Quantum Probability and Applications IV. Proceedings, 1987. VI, 355 pages. 1989.

Vol. 1397: P.R. Turner (Ed.), Numerical Analysis and Parallel Processing. Seminar, 1987. VI, 264 pages. 1989.

Vol. 1398: A.C. Kim, B.H. Neumann (Eds.), Groups – Korea 1988. Proceedings. V, 189 pages. 1989.

Vol. 1399: W.-P. Barth, H. Lange (Eds.), Arithmetic of Complex Manifolds. Proceedings, 1988. V, 171 pages. 1989.

Vol. 1400: U. Jannsen. Mixed Motives and Algebraic K-Theory. XIII, 246 pages. 1990.

Vol. 1401: J. Steprāns, S. Watson (Eds.), Set Theory and its Applications. Proceedings, 1987. V, 227 pages. 1989.

Vol. 1402: C. Carasso, P. Charrier, B. Hanouzet, J.-L. Joly (Eds.), Nonlinear Hyperbolic Problems. Proceedings, 1988. V, 249 pages. 1989.

Vol. 1403: B. Simeone (Ed.), Combinatorial Optimization. Seminar, 1986. V, 314 pages. 1989.

Vol. 1404: M.-P. Malliavin (Ed.), Séminaire d'Algèbre Paul Dubreil et Marie-Paul Malliavin. Proceedings, 1987 – 1988. IV, 410 pages. 1989.

Vol. 1405: S. Dolecki (Ed.), Optimization. Proceedings, 1988. V, 223 pages. 1989.

Vol. 1406: L. Jacobsen (Ed.), Analytic Theory of Continued Fractions III. Proceedings, 1988. VI, 142 pages. 1989.

Vol. 1407: W. Pohlers, Proof Theory. VI, 213 pages. 1989.

Vol. 1408: W. Lück, Transformation Groups and Algebraic K-Theory. XII, 443 pages. 1989.

Vol. 1409: E. Hairer, Ch. Lubich, M. Roche. The Numerical Solution of Differential-Algebraic Systems by Runge-Kutta Methods. VII, 139 pages. 1989.

Vol. 1410: F.J. Carreras, O. Gil-Medrano, A.M. Naveira (Eds.), Differential Geometry. Proceedings, 1988. V, 308 pages. 1989.

Vol. 1411: B. Jiang (Ed.), Topological Fixed Point Theory and Applications. Proceedings, 1988. VI, 203 pages. 1989.

Vol. 1412: V.V. Kalashnikov, V.M. Zolotarev (Eds.), Stability Problems for Stochastic Models. Proceedings, 1987. X, 380 pages. 1989.

Vol. 1413: S. Wright, Uniqueness of the Injective III_1 Factor. III, 108 pages. 1989.

Vol. 1414: E. Ramírez de Arellano (Ed.), Algebraic Geometry and Complex Analysis. Proceedings, 1987. VI, 180 pages. 1989.

Vol. 1415: M. Langevin, M. Waldschmidt (Eds.), Cinquante Ans de Polynômes. Fifty Years of Polynomials. Proceedings, 1988. IX, 235 pages. 1990.

Vol. 1416: C. Albert (Ed.), Géométrie Symplectique et Mécanique. V, 289 pages. 1990.

Vol. 1417: A.J. Sommese, A. Biancofiore, E.L. Livorni (Eds.). Algebraic Geometry. Proceedings, 1988. V, 320 pages. 1990.

Vol. 1418: M. Mimura, Homotopy Theory and Related Topics. Proceedings, 1988. V, 241 pages. 1990.

Vol. 1419: P.S. Bullen, P.Y. Lee, J.L. Mawhin, P. Muldowney, W.F. Pfeffer (Eds.), New Integrals. Proceedings, 1988. V, 202 pages. 1990.

Vol. 1420: M. Galbiati, A. Tognoli (Eds.), Real Analytic Geometry. Proceedings, 1988. IV, 366 pages. 1990.

Vol. 1421: H.A. Biagioni, A Nonlinear Theory of Generalized Functions. XII, 214 pages. 1990.

Vol. 1422: V. Villani (Ed.), Complex Geometry and Analysis. Proceedings, 1988. V, 109 pages. 1990.